I0482659

Disclaimer

Book Title: PERFORMANCE OF NEW AND AGED RESIDENTIAL FIRE SMOKE DETECTORS

Book Author: Jason D. Averill; Richard G. Gann; William F. Guthrie; Daniel Murphy;

Book Abstract: As part of the Consumer Product Safety Commission (CPSC) technical staff program to determine the effects of emissions from imported drywall on residential electrical, gas distribution, and fire safety components, the National Institute of Standards and Technology (NIST) has generated data to aid assessment of whether there has been a loss of functionality of fire smoke detectors exposed to those emissions. NIST tested four sets of smoke detectors in the Fire Emulator/Detector Evaluator (FEDE) and the UL 217 test apparatus. Set 1 (ionization only) detectors were provided by CPSC staff and described as having been installed in homes with imported drywall; Set 2 (ionization only) detectors (same models as Set 1 but different batches) were described as coming from contemporaneous homes without the presence of imported drywall; Set 3 smoke detectors were purchased new by NIST and tested as received; and Set 4 comprised detectors from Set 3 that had been subjected to an accelerated aging protocol to simulate 10-year exposure to the effluent from imported drywall. Detectors from Set 1 and Set 2 activated within UL 217 sensitivity test parameters. Tests using the FEDE found differences in sensitivity between Sets 1 and 2 and between the ionization detectors in Sets 3 and 4 that were numerically small compared to the allowable performance range. An even smaller improvement in sensitivity was found for photoelectric smoke detectors between Sets 3 and 4. Two of the Set 1 smoke detectors sensed the presence of smoke, but would not send the signal to activate interconnected alarms. Two other Set 1 smoke detectors failed to operate under AC power, but operated properly under the required 9V battery back-up. It could not be determined whether the observations of Set 1 detector performance could be attributed to exposure to the emissions from imported drywall or to other factors.

Citation: NIST TN - 1691

Keywords: drywall; fire; fire smoke; ionization smoke detectors; photoelectric smoke detectors; UL 217

NIST Technical Note 1691

PERFORMANCE OF NEW AND AGED RESIDENTIAL FIRE SMOKE ALARMS

Jason D. Averill

Richard G. Gann

Daniel C. Murphy

Fire Research Division
Engineering Laboratory

William F. Guthrie

Statistical Engineering Division
Information Technology Laboratory

U.S. Department of Commerce

Gary Locke, Secretary

Patrick D. Gallagher

Under Secretary of Commerce for Standards and Technology
Director, National Institute of Standards and Technology

ii

National Institute of Standards and Technology Technical Note 1691

Natl. Inst. Stand. Technol. Tech. Note 1691, 106 pages (April 2011)

CODEN: NTNOEF

Technical Report to the U.S. Consumer Product Safety Commission

CPSC - I - 09 - 0024

Rohit Khanna, Fire Protection Engineer

This report was prepared for the Commission pursuant to Interagency Agreement CPSC-I-09-0024. It has not been reviewed or approved by, and may not necessarily reflect the views of, the Commission.

This page intentionally left blank.

ABSTRACT

The U.S. Consumer Product Safety Commission (CPSC) initiated a program to determine the effects of emissions from problem drywall on residential electrical, gas distribution, and fire safety components. As part of this program, the National Institute of Standards and Technology (NIST) generated data to aid assessment of whether there has been a loss of functionality of residential fire smoke alarms exposed to those emissions. NIST tested four sets of smoke alarms in its Fire Emulator/Detector Evaluator (FEDE). Set 1 alarms were provided by CPSC staff and described as having been installed in homes with problem drywall; Set 2 alarms (same models as Set 1, but different batches) were described as coming from contemporaneous homes without the presence of problem drywall; Set 3 smoke alarms were purchased new by NIST and tested as received; and Set 4 comprised smoke alarms from Set 3 after they had been subjected to an accelerated aging protocol, the Battelle Class IV corrosivity environment. Sets 1 and 2 were also tested in a UL 217 test apparatus. Alarms from Set 1 and Set 2 were activated within UL 217 sensitivity test parameters. Tests using the FEDE found differences in sensitivity between Sets 1 and 2 and between the ionization alarms in Sets 3 and 4 that were numerically small compared to the allowable performance range. An even smaller difference in sensitivity was found for one of three models of photoelectric smoke alarms between Sets 3 and 4. NIST observed that two of the 29 Set 1 smoke alarms sensed the presence of smoke but did not send a signal to activate interconnected alarms. Two other Set 1 smoke alarms failed to operate under 110 VAC power but operated properly under the required 9 VDC battery back-up. Neither type of failure is associated with the smoke sensing mechanism.

Keywords: drywall; fire; fire detection; fire smoke; ionization smoke alarms; photoelectric smoke alarms; UL 217.

This page intentionally left blank.

SUMMARY

U.S. Consumer Product Safety Commission (CPSC) technical staff is conducting an engineering test program to determine the effects of emissions from problem[i] drywall on residential electrical, gas distribution, and fire safety components as part of its overall investigation. For this program, the National Institute of Standards and Technology (NIST) has conducted testing to aid assessment of whether there has been a loss of functionality of residential fire smoke alarms (hereafter referred to as "smoke alarms") and automatic fire sprinklers exposed to those emissions. This document presents the results of the NIST testing regarding smoke alarms.

The testing was designed to address three questions:

1. Do smoke alarms exposed to the effluent from problem drywall activate within allowable tolerances?

2. Do smoke alarms exposed to the effluent from homes with problem drywall perform differently from similar vintage smoke alarms from homes without problem drywall?

3. Does a 10-year exposure to a Battelle Class IV corrosivity environment, simulated through an accelerated aging test protocol, affect smoke alarm performance?

NIST conducted tests of 79 smoke alarms in the Fire Emulator/Detector Evaluator (FEDE) and tests of 24 smoke alarms in an Underwriters Laboratories (UL) 217 test apparatus, the *de facto* national standard. The smoke alarms were tested in four sets:

(Set 1) 29 smoke alarms, provided by CPSC staff and described as having been installed in homes constructed around 2006 with problem drywall, and removed in 2009;

(Set 2) 14 smoke alarms, provided by CPSC staff and described as being from homes constructed around 2006, but without problem drywall, and removed in 2009;

(Set 3) 36 smoke alarms purchased by NIST in 2009, tested as received; and

(Set 4) the same 36 smoke alarms from Set 3 that had been subjected to a concentration of corrosive gases designed to simulate 10-year exposure through an accelerated aging protocol.

All of the Set 1 and Set 2 smoke alarms operated on the ionization principle because that was the only type of smoke alarm that could be obtained from homes of the era and in the locales where problem drywall had been used. Equal numbers of the store-bought smoke alarms in Sets 3 and 4 operated on ionization and photoelectric principles.

The smoke alarms from Set 1 and Set 2 activated within UL 217 sensitivity test parameters. Comparing the sensitivity of the Set 1 smoke alarms relative to the same models in Set 2, tests using the FEDE found a reduction in sensitivity that was small compared to the allowable performance range. A reduction in sensitivity that was small compared to the allowable range was also found for all three models of the Set 4 ionization smoke alarms, relative to the sensitivity of the same smoke alarms tested as Set 3. An even smaller change in sensitivity between Set 3 and Set 4 testing was found for one of the three models of photoelectric smoke alarms. All of the changes in sensitivity were small compared to the allowed performance range.

---

[i] CPSC uses the term "problem drywall" to refer to drywall associated with corrosion of metal in homes. This has also been reported as "Chinese drywall," but CPSC staff has found that not all Chinese or imported drywall is problem drywall, and has not ruled out the possibility that some domestic drywall could be problem drywall.

In addition, two of the 29 Set 1 smoke alarms sensed the presence of smoke, produced an audible signal, but would not send a voltage signal needed to activate interconnected smoke alarms, which typically would be located in other parts of the home. Two other Set 1 smoke alarms failed to operate under AC power but operated properly under the required 9 V back-up battery. None of the other tested smoke alarms exhibited these problems. Neither type of failure (not sending a signal to interconnected smoke alarms or failing to operate under AC power) is associated with the smoke sensing mechanism.

Table of Contents

I. OBJECTIVE AND BACKGROUND

At the request of the U.S. Consumer Product Safety Commission (CPSC), the National Institute of Standards and Technology (NIST) tested and analyzed the performance of residential fire smoke alarms to aid the CPSC in ascertaining possible changes in functionality and resulting fire safety risk arising from smoke alarm exposure to emissions from problem drywall. This Technical Note documents the NIST research.

Some problem drywall installed in U.S. homes has been reported to be associated with corrosion to central air conditioning components, copper tubing, and exposed copper wiring. [1] There have been reports of premature failures of various household items: air conditioner evaporator coils, electric appliances, televisions, and electrical switches in homes. A range of health issues have been reported by residents in these homes as well. [2]

The CPSC's investigation into the possible effects of the problem drywall includes three tracks: (1) evaluating the relationship between the drywall and reported health issues; (2) evaluating the relationship between the drywall and possible effects on electrical, gas distribution, and fire safety components that could result in potential fire and shock hazards; and (3) tracing the origin and distribution of the drywall in commerce to identify the scope of potential problems presented by the drywall.

As part of this investigation, the CPSC and the U.S. Environmental Protection Agency (EPA) and its partners conducted the following studies to determine the differences in composition between problem drywall and domestic drywall typically used:

- In-home air sampling studies, including continuous, real-time measurements of sulfur compounds, acids, and other gases. The sampling extended over days because many health symptoms reportedly occurred after hours of sleeping. [1]

- Laboratory elemental analysis of domestic and problem drywall, with characterization of any differences.

- Laboratory chamber studies of domestic and problem drywall to separate and isolate chemical emissions from drywall as opposed to chemicals emitted from other home products (*e.g.,* carpets, cleaners, paint, adhesives, and beauty products).

These studies have enabled the agencies to identify contaminated drywall and to characterize differences in the chemical composition of the emitted gases between problem drywall and domestic drywall. [3]

A reference environment has been established for replicating the effects of the emissions on metallic and electrical components and, by increasing the temperature and concentration of the contaminating gases, accelerating the aging of the components. [4] This environment, designated Battelle Level IV, is an accelerated aging protocol designed to replicate the outcome of 10 years of exposure in two weeks. [5]

This report describes the operating principles of the smoke alarms, identifies the alarms included in the NIST evaluation, details the apparatus and procedures used to measure alarm performance, and presents the test data. The report concludes with an analysis of the test results for requirements for smoke alarm sensitivity, as well as the degree of change in sensitivity as a result of exposure to contaminated atmospheres.

This page intentionally left blank.

II. SMOKE ALARM ACTIVATION PRINCIPLES

A. COMBUSTION PRODUCTS

During a house fire, combustion products are emitted from the burning object(s). The combustion products contain carbonaceous soot particles and liquid aerosols (hereafter referred to collectively as "smoke"), as well as toxic gases, including carbon monoxide and corrosive or asphyxiant gases, depending on the chemistry of the underlying product consumed. Residential building and fire codes require fire smoke alarms to be installed in strategic locations throughout the home to sense the presence of an unwanted fire rapidly and accurately and to produce an audible alarm, which notifies the occupants of the presence of the fire. Typically, these smoke alarms are required to meet the Underwriter's Laboratories (UL) 217 [6] test requirements to be deemed acceptable for installation in residences.

Residential smoke alarms are commonly based on one of two operating principles: ionization and photoelectric. The former dominates current usage. In both designs, room air enters a sensing chamber in the smoke alarm. When the presence of smoke in the air within the chamber exceeds a threshold level, the unit emits a loud alarm.

B. IONIZATION SMOKE ALARMS

Ionization smoke alarms contain a small mass of Americium (^{241}Am). With a radioactive half-life of 432 years, this metal emits a steady stream of positively charged alpha particles. Power is supplied to positive and negative electrodes in the sensing chamber. The alpha particles ionize the air molecules in the chamber, resulting in equal numbers of negatively charged electrons and positively charged molecular ions. The electrons and molecular ions flow rapidly to their respective electrodes, and the resulting electrical current is monitored. When smoke particles enter the chamber, the charged species attach to them. The charged smoke particles are far larger than the free electrons and molecular ions and move more slowly to the electrodes. The measured property is the resulting decrease in electrical current.

C. PHOTOELECTRIC SMOKE ALARMS

Photoelectric smoke alarms use the light-scattering properties of smoke particles and aerosols as the sensing mechanism. A light-emitting diode (LED) source transmits a light beam across the sensing chamber. An off-axis light sensor, placed at an angle to the light beam, is also located within the sensing chamber. In the absence of smoke, the light beam passes in a straight line (on axis) through the chamber; and the light sensor does not receive any light. When smoke enters the sensing chamber, some of it crosses the path of the light beam. Light is scattered off the particles and droplets in all directions, and some of that light reaches the off-axis light sensor. The measured property is the intensity of light reaching the sensor.

3

This page intentionally left blank.

III. EXPERIMENTAL PLAN

A. GENERAL FORMULATION

The experimental plan was designed to address three questions:

1. Do smoke alarms exposed to the emissions from problem drywall activate within allowable tolerances?

2. Do smoke alarms exposed to emissions from problem drywall perform differently from similar vintage smoke alarms in homes without problem drywall?

3. Does a 10-year exposure, simulated by a two-week exposure to the Battelle Class IV corrosivity environment using an accelerated aging test protocol, affect smoke alarm performance?[ii]

To answer these questions, the exposure history of the smoke alarm was a critical consideration. Thus, the plan included testing of alarms from four exposure conditions:

(Set 1) 29 smoke alarms provided by CPSC staff and described as having been installed in homes constructed with problem drywall around 2006 and removed in 2009;

(Set 2) 14 smoke alarms, also provided by CPSC staff and described as having been installed in homes constructed without problem drywall around 2006 and removed in 2009;

(Set 3) 36 smoke alarms purchased by NIST in 2009, tested as received; and

(Set 4) the same 36 smoke alarms that after testing as Set 3, had been subjected to a concentration of Battelle Level IV corrosive gases, designed to simulate 10-year exposure through an accelerated aging procedure.

Additional factors were included in the experimental plan:

- Sets 3 and 4 included both ionization and photoelectric units. The only smoke alarms available for Sets 1 and 2 were ionization units.

- To assess the impact of any product design differences, devices from more than one manufacturer were tested, where possible.

- Replicate tests were conducted to enable estimation of experimental repeatability.

- The smoke alarms were examined for their sensitivity to smoke from smoldering cotton wicks because this is the standard source used in UL 217.

Tests were conducted in two laboratories using two different apparatus: the Fire Emulator/Detector Evaluator (FEDE) at NIST and a UL 217 test apparatus (the *de facto* national certification test standard), with the UL 217 tests conducted in a certified third party laboratory. Descriptions of the principles of operation of the two apparatuses appear in Chapters IV and V, respectively, along with an explanation of the purpose for which each was employed.

[ii] NIST did not investigate the appropriateness of the composition of this particular environment, nor did NIST investigate whether two weeks of exposure in this environment was equivalent to 10 years of exposure of home smoke alarms to the emissions from contaminated drywall.

Nine different models of smoke alarms were tested. Under ideal circumstances, the same model of smoke alarm from the same manufacturing batch would be evaluated as part of all four Sets. However, in the time between smoke alarm installation in homes with problem drywall (generally 2006), and the time that smoke alarm evaluation was initiated at NIST (2009), smoke alarm designs had changed, and the exact models could no longer be purchased as new units. Further, CPSC staff provided no photoelectric smoke alarms from the problem drywall or domestic drywall homes, as none were found in the sampled houses.

The smoke alarm models that populate the four Sets listed in Section III.A of this Technical Note are further described as follows:

1. CPSC staff provided 29 smoke alarms described as being from homes constructed with problem drywall around 2006, comprising three models (denoted as A, B, and C in this report). All 29 units were ionization models.

2. CPSC staff provided 14 smoke alarms described as being from homes constructed with domestic drywall at roughly the same time as the homes with problem drywall, comprising two of the same models that were found in contaminated homes (denoted A and B in this report).[iii] All 14 units were ionization models.

3. NIST procured from common commercial sources 36 new ionization smoke alarms, representing three different models (denoted D, E, and F in this report).[iv] NIST also procured from common commercial sources 36 new photoelectric smoke alarms, representing three different models (denoted G, H, and I in this report). For both the ionization and photoelectric smoke alarms, six units of each model (for a total of 36) were tested in the FEDE as received.

4. For both the ionization and photoelectric smoke alarms, these same 36 units were tested in the FEDE after exposure to an accelerated aging protocol at Sandia National Laboratories.

A summary of the experimental test matrix is shown in Table 1.

[iii] For an ideal comparison of the effects of exposure to problem drywall and exposure to domestic drywall, all the other exposure conditions would have been identical. These include criteria such as the two sets of houses being otherwise identical, the two sets of smoke alarms being from the same manufacturers' batches, the indoor contaminant levels in the two sets of homes being otherwise identical, both sets of smoke alarms having been installed at the same time and in similar locations. Determining any of these criteria was not possible in this study. This limited, but did not eliminate, the ability to derive information regarding the effects of exposure to the effluent from problem drywall.

[iv] To enable comparison with the smoke alarms in Sets 1 and 2, it would have been preferable for the smoke alarms in Sets 3 and 4 to be at least the same models. However, conventional retail practice is not to sell units that are more than one year old. In 2009, the alarms in Sets 1 and 2 were approximately three years old. Thus, it was not possible to purchase new units from the same batches. Furthermore, in the ensuing three years, (a) new (albeit similar) models had replaced the original models, and (b) the labeled sensitivities of the new Set 3 models all differed significantly from the original Set 1 and Set 2 models. This limited, but did not eliminate, the ability to derive information regarding the effect of nominal three-year exposures in Sets 1 and 2 and the simulated 10-year exposure in Set 4.

	Exposure Condition	Smoke Alarm Type	Model	Number of Units	Number of FEDE Tests	Number of UL217 Tests
Set 1	Three-year Exposure to Problem Drywall	Ionization	A	15	45	12
			B	7	21	12
			C	7	24	12
Set 2	Three-year Exposure to Typical Residence Without Problem Drywall	Ionization	A	7	22	-
			B	7	22	-
Set 3	New Smoke Alarms From Commercial Sources	Ionization	D	6	18	-
			E	6	18	-
			F	6	18	-
		Photoelectric	G	6	18	-
			H	6	20	-
			I	6	18	-
Set 4	New Smoke Alarms With Accelerated 10-year Exposure to Battelle Class IV Corrosion (Accelerated Aging)	Ionization	D	6	19	-
			E	6	18	-
			F	6	18	-
		Photoelectric	G	6	18	-
			H	6	18	-
			I	6	18	-

TABLE 1: TEST MATRIX

C. ACCELERATED AGING PROTOCOL

Staff at the Sandia National Laboratories conducted accelerated aging tests in their Facility for Atmospheric Corrosion Testing (FACT II, Figure 1). Observations from metal coupons placed in homes constructed with problem drywall[2] qualitatively matched Battelle Class IV corrosion levels, a corrosion process dominated by sulfide creep.[4] Smoke alarms were exposed for two weeks, which roughly represented a field exposure in a light industrial environment of 10 years. The Class IV environment contains H_2S, NO_2, Cl_2, and water vapor. The nominal composition of the inflow was:

- 200 nL/L H_2S
- 200 nL/L NO_2
- 50 nL/L Cl_2
- 75 % relative humidity

The pollutant gasses were supplied using permeation tubes, and mass flow controllers were used to maintain flow. The chamber volume was 300 L, and the total flow was set at 12 L/min. The gas temperature and pressure were nominally 50 °C and 101 kPa, respectively. A schematic of the system is shown in Figure 1. Within the reaction chamber, all smoke alarms were oriented in the same direction to ensure that entry characteristics for the gas mixture were consistent.

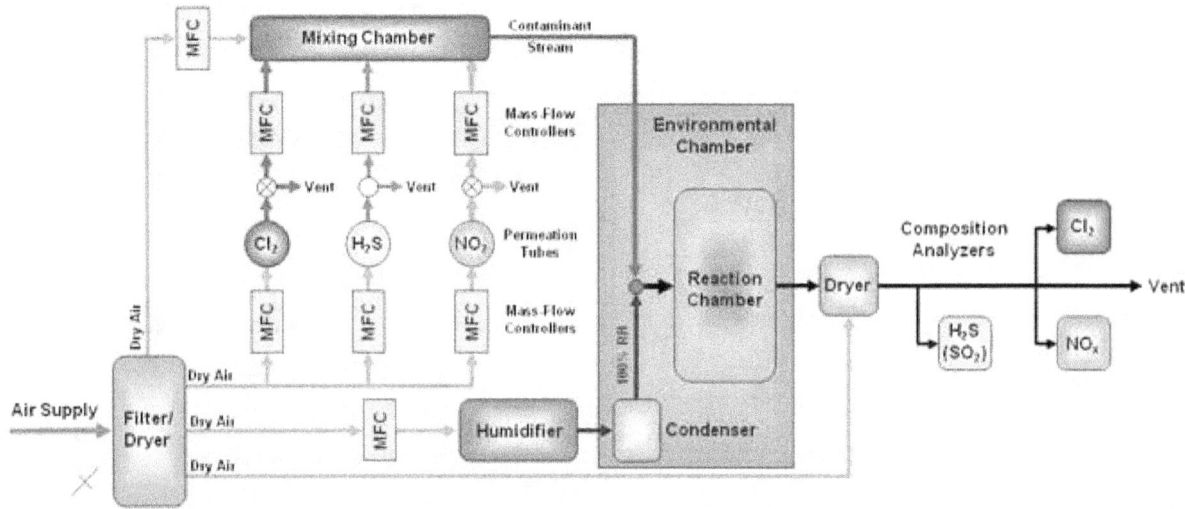

FIGURE 1: FACT II SCHEMATIC

(Figure Courtesy of Sandia National Laboratories, used with permission)

A. GENERAL DESCRIPTION

Testing of all smoke alarms was conducted using the Fire Emulator/Detector Evaluator (FEDE) developed at NIST. [7,8] The FEDE is shown schematically in Figure 2 and photographically in Figure 3. It is a single-pass wind tunnel designed to generate gas flows capable of recreating a wide range of environmental conditions for the purpose of testing a variety of fire sensors. The following section briefly describes the relevant features and any modifications to the apparatus and operating procedure.

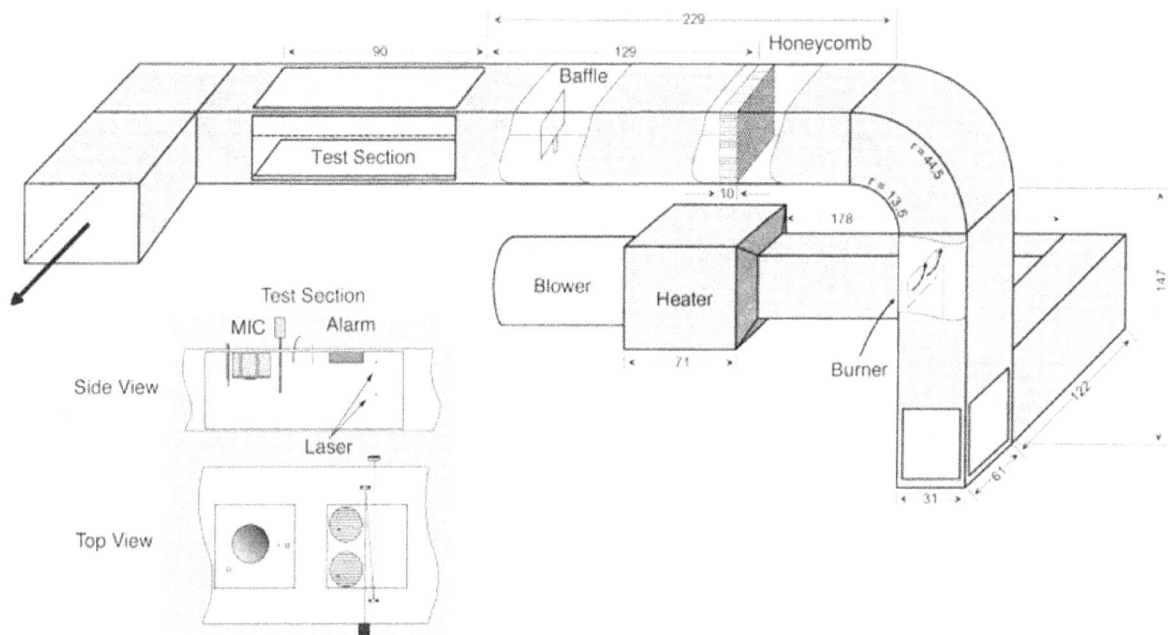

FIGURE 2: SCHEMATIC OF THE FIRE EMULATOR/DETECTOR EVALUATOR

FIGURE3: PHOTOGRAPH OF THE FIRE EMULATOR/DETECTOR EVALUATOR
Note: The red arrow denotes the test section and the yellow arrow denotes the smoke source.

The FEDE is composed primarily of 610 mm x 310 mm (internal dimension), polytetrafluoroethylene (PTFE)-coated metal ductwork that draws in ambient air and exhausts directly into a laboratory ventilation hood. For determining the ordinary sensitivity of smoke alarms, the only parameters directly controlled were flow velocity and smoke concentration. The air velocity was controlled by means of a variable speed fan at the entrance of the FEDE duct, as seen in Figure 2. Because of the different sensitivity ranges of these photoelectric and ionization smoke alarms, the smoke concentrations that led to activation were achieved using two different flow velocities: nominally 0.2 m/s for the former and 0.3 m/s for the latter. The temperature was nominally 21 °C.

The device used to generate the smoldering combustion products is located at the base of the vertical riser (see the arrow in the lower right corner of Figure 3), within a short chimney to prevent the flow within the duct from impinging directly upon the burning fuel. For each test, this device suspended 12 cotton wicks vertically, with their bottom ends in contact with coils of high resistivity electrical wire, as shown in Figure 5. A computer signal initiated ignition of a bank of one to four wicks at a time, with the smoke from each successive bank adding to the smoke level generated by the previously ignited wicks (see Section IV.B). After the start of the test, one bank of wicks was ignited approximately every 163 s (see Section IV.C).

The smoke-laden flow continued through the ductwork via a honeycomb flow straightener into the test section, shown in Figure 4. The test section contains mounting points and electrical connections for the smoke alarms, as well as equipment used to monitor and characterize the environment within the FEDE.

FIGURE 4: CLOSE-UP OF THE TEST SECTION, SHOWING TWO CEILING-MOUNTED SMOKE ALARMS AND TWO LASERS FOR SMOKE OBSCURATION MEASUREMENT

FIGURE 5: CLOSE-UP OF 12-WICK SMOKE SOURCE. LEFT: PRIOR TO A TEST; RIGHT: DURING A TEST

Characterization of the smoke during testing was achieved using a measuring ionization chamber (MIC) and by extinction of laser light through the test section. The MIC, essentially a reference ionization detector, was situated 310 mm downstream of the smoke alarms being tested. The MIC is a standard device for defining alarm limits of ionization alarms and is referenced in UL 217 and international standards. The ionizing radioactive material is the same ^{241}Am used in ionization smoke alarms, but is present at approximately three times the mass. The material is contained in a

11

chamber, like a smoke alarm source. The output current was measured using a precision picoammeter, and values are expressed in terms of picoamperes (pA).

Two pairs of lasers and photodetectors were located within the duct, 73 mm upstream of the centerline of the smoke alarms at heights of 152 mm and 252 mm. Both the lasers and the photodetectors were located outside of the transparent test section enclosure. The beam of each laser was reflected twice by mirrors within the test section in order to increase the path length to 1.52 m. The reported parameter was the laser transmittance, the ratio of the light intensity at a photodetector (I) divided by the light intensity when there was no smoke in the system (I_o).

The only signal from the smoke alarm being tested was the loud noise upon activation. The characterization of the smoke at that time was from the MIC value and/or the laser extinction value.

B. TEST PROTOCOL

An important consideration in designing a test series to discern the presence or absence of changes in properties (here, the sensitivity of the smoke alarms) is the isolation of test factors that might confound that determination. In this study, the following were varied in most of the testing:

- Test order. This included replicates of tests for a particular smoke alarm, tests of units of the same model of smoke alarm, and tests of different models of smoke alarms. Of course, the tests of all the smoke alarms sent for accelerated aging were conducted before the smoke alarms were sent, and all the smoke alarms that had been through the accelerated aging protocol were tested when returned to NIST.

- Day of replicate tests. In order to avoid "memory effects," such as a long delay in a smoke alarm returning to normal function after activation, replicate testing of individual ionization units was performed on different days. This also enabled identification of any drift in the test apparatus that might have affected test results.

- Position (front or rear) of the ionization smoke alarm in the FEDE. This enabled identification of any effect on sensitivity that might have resulted (*e.g.*, from a difference in the smoke entrance into the sensing chamber).

With these provisions, the conduct of the tests was as follows: As shown in Figure 4, for each test, two smoke alarms of the same operating principle (ionization or photoelectric) were placed in the test section and located at the same downstream distance to ensure they did not affect each other.

After the airflow was set, the approximate sensitivity of the pair of alarms was determined. The wick igniter was then loaded such that an alarm could be activated by the additional smoke from a single wick. For example, if an alarm was expected to be activated by the smoke load produced by 11 wicks, the wick igniter was loaded with four wicks in the first two banks and only one wick in each remaining bank. The advantage of this method is that it reduces the difference in smoke concentration between the pre- and post-activation signals, improving the precision with which the sensitivity of the alarm can be determined.

Activation of each alarm was recorded by monitoring the electronic interconnect signal[v] directly from the smoke alarm. One brand of smoke alarm produced a nonstandard interconnect signal. For this brand, a manual switchbox was used to generate a standard interconnect signal to the data acquisition system.

C. DATA REDUCTION

The staged smoldering wick ignition method used to generate smoke in the FEDE produces a unique pattern of increasing smoke concentrations, which necessitates the use of a specific method of data reduction in order to quantify the sensitivity of a smoke alarm. As each bank of up to four wicks is ignited, there is a brief surge of smoke production before settling to a nominally constant rate. Figure 6 shows the obscuration of one of the laser beams and the MIC reading during a typical test. Each sharp decrease marks the ignition of a bank of wicks. The subsequent quasi-steady "plateau" regions, some of which are indicated by red circles in the figure, are representative of the smoke concentration due to the steady smoldering of all presently involved wicks.

The interpretation of the MIC and laser measurements relative to smoke alarm sensitivity is further complicated by differences in smoke transport, especially during periods of rapidly changing smoke concentration. The laser obscuration measurements are effectively instantaneous representations of the test section environment. By contrast, both the MIC and the commercial smoke alarms being examined can only measure changes as quickly as smoke can pass through their outer casing and fill the sensing chamber. [9] In addition, there are likely to be differences in these flushing times among different models of smoke alarms and the MIC.

Both of these considerations are factored into the test procedure by setting the time between ignitions of successive banks to be sufficiently long for these steady states to be realized for all the devices. A value and uncertainty for a steady state is determined by averaging over the observed fluctuations.

A sensitivity determination then proceeds as follows: As successive banks of wicks are ignited, the initial smoke transient from one bank will activate a smoke alarm. As this transient subsides, the alarm state may cease. The next bank of wicks will again lead to an alarm. Relaxation may follow this as well. At some point, one of the transients will initiate an alarm that does not subside. There are then two ways of extracting a current or optical density to characterize the smoke alarm sensitivity:

1. Use the plateau value for the first steady state that sustains the smoke alarm in a state of alarm. This is slightly conservative. The uncertainty range in this value is "plus the difference between the current plateau value and the next higher obscuration plateau value" and "minus zero." Added to this is the uncertainty in determining the plateau values.

2. Use the average of the two last plateau values (*i.e.*, the one that did not sustain a state of alarm and the one that did sustain that state). The uncertainty in this value is ± (one-half the difference between the two values) plus the uncertainty in determining the plateau values.

---

[v] An interconnect signal activates other (connected) smoke alarms in the residence in order to increase the likelihood that the alarm signal will result in occupant awareness of the fire, particularly for occupants remote from the room of fire origin. Interconnection of smoke alarms is not required in all jurisdictions.

For this project, the first approach was used.

For the MIC measurements, the standard deviation among the data points used to determine the plateau value was typically 1.2 pA. The difference between adjacent plateaus was typically 2 pA. (A "good" test was defined by the final two plateaus being the smoke from a single additional wick. A typical value for the MIC current for clean air was near 100 pA.) For the laser transmission measurements, the standard deviation among the I/I_0 data points used to determine the plateau value was typically 0.01. The difference in I/I_0 between adjacent plateaus was typically 0.03.

The final step in the analysis was a normalization procedure. The MIC or laser signal may drift slightly between or during a test. Each MIC ionization activation value was normalized to the initial MIC output because the intratest drift was negligible. The sensitivity determined using the laser photodetectors was adjusted to account for both initialization and intratest drift.

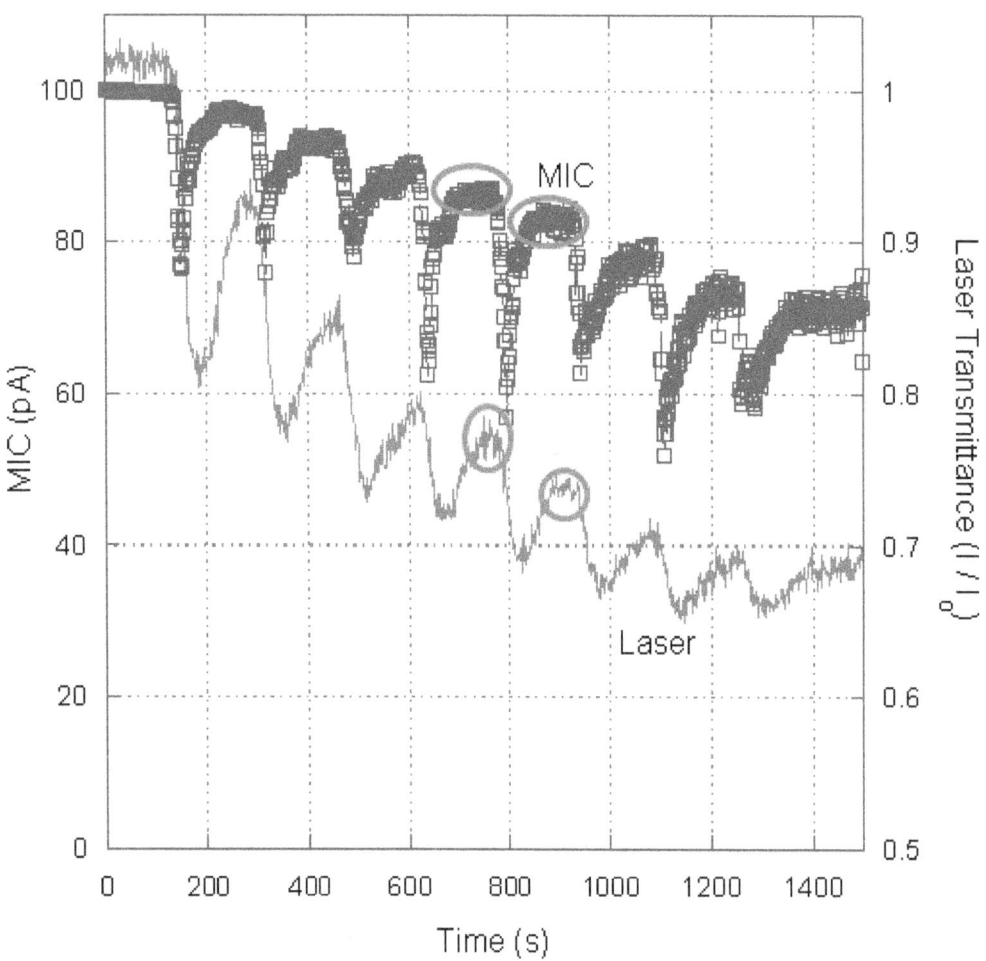

FIGURE 6: MIC AND LASER RESPONSE FOR A TYPICAL FEDE TEST

A. PURPOSE

The first task for this research project (Section III.A) was to determine the extent to which the Set 1 smoke alarms activated within the parameters allowed by the *de facto* national certification test standard, UL 217. To obtain the necessary data for this determination, NIST transported smoke alarms from Set 1 and Set 2 to an independent test laboratory that routinely performs the UL 217 protocol. The UL 217 test apparatus is a recirculating flow loop device that uses smoldering cotton wicks (the same as those in the FEDE) within the flow path to generate smoke particles. The concentration of these particles increases over time. When the smoke alarm activates, the test is terminated. With NIST staff present, the laboratory staff conducted 36 tests of 12 smoke alarms (*i.e.*, each unit was tested in triplicate).

B. APPARATUS DESCRIPTION

The UL 217 apparatus is designed to generate a "homogeneous aerosol mix and a laminar air flow across the alarm, adjustable from 0.16 to 0.76 m/s."[6] In the reference configuration, the inside duct axial dimension is 1.67 m long, with a cross sectional area of 490 mm x 460 mm wide. A variable speed fan is used to generate flow within the duct. A honeycomb-style flow straightener is used to reduce the level of turbulence.

To measure smoke obscuration, a photovoltaic cell, mounted to limit the detection of forward-scattered light, receives white light from a low-voltage automotive spotlight shone over a 1.52 m path length. To measure ionization response, a MIC is mounted adjacent to the alarms, which are located roughly in the center of the flow path. According to UL 217, the uncertainty in the MIC current at the time of alarm was typically 1.2 pA. (A typical value for the MIC current for clean air was near 100 pA.)

C. TEST PROTOCOL

As prescribed in the UL 217 document, the tests were conducted at an ambient temperature of (23 ±3) °C and at a relative humidity between 30 % and 70 %.[vi] The air velocity in the test compartment was maintained at (0.16 ± 0.001) m/s. The smoke was admitted into the test chamber, and operation was continued until the alarm was activated. When the tested unit went into alarm, the obscuration, MIC reading, and time were recorded.

The test chamber was flushed with clean air following a test until the MIC and light beam indicated a clear condition. The air flow was then allowed to stabilize for at least 30 s before each test trial.

According to UL 217, smoke alarms are to activate when the light extinction, as a result of exposure to the smoke from cotton wicks, is within a range of obscuration of 1.6 %/m to 12.5 %/m or the MIC current is between 93 pA and 37.5 pA. For the 1.52 m light path in the UL 217 apparatus, this light obscuration corresponds to a light transmittance range of approximately 0.81 to 0.98.

[vi] This allowed range is quite wide and might have resulted in variation of aerosol aggregation. However, the need was to perform the test and obtain test results as prescribed in the Standard.

The UL 217 test standard differs from the FEDE test protocol in the following important ways:

1. The flow velocity in the FEDE is higher than the UL 217 standard. This can have an effect on the time needed for the smoke-laden airflow to flush through the sensing chamber of the smoke alarm being tested.

2. Smoke is recirculated in the UL 217 test method (flow loop), while the FEDE is a single-pass flow tunnel. Combined with the slower flow velocity, this means that there is far more time for the smoke to age in the UL 217 test method. Aging includes the agglomeration of soot to form fewer, larger particles and the coalescing of aerosols to form larger droplets. The largest particles and droplets may not follow the flow streamlines and thus could stick to surfaces or settle due to gravity. Both the light obscuration and the electric current reduction in the MIC have a dependence on the particle size distribution of the smoke. A shift to larger particles from aging of cotton smolder smoke tends to increase the obscuration and decrease the MIC current. The FEDE cotton smolder smoke light obscuration and MIC output levels are within the range allowed in UL 217.

3. The FEDE provides significantly greater flexibility for testing smoke alarms, preserving options for further investigation of any questions that might have arisen during testing. These options included the use of multiple particle sources, a greater range of flow characteristics, and additional measurement capacity.

There was an advantage to testing all the smoke alarms using a single apparatus. It was anticipated that some comparisons would be made among all four Sets of measurements; and, due to reasons given in the first two bullets above, it is difficult to relate quantitatively the results from the two apparatuses. The FEDE was used for this purpose because, as noted in the third bullet above, the FEDE provided the capability to conduct additional tests under varied conditions if more information might help address the three research questions more thoroughly.

However, the UL 217 apparatus is the *de facto* national standard for qualifying smoke alarms. Therefore, NIST also developed data on the Set 1 and Set 2 smoke alarms, using UL 217 to answer the first research question, which addresses whether smoke alarms exposed to the emissions from problem drywall activate within allowable tolerances.

VI. RESULTS AND ANALYSIS

A. LIMITATIONS

The findings of this report are subject to certain limitations that should be considered when interpreting the data or findings.

Only 43 Set 1 and Set 2 residential fire smoke alarms from homes were available for analysis. The findings cannot be generalized to the universe of all smoke alarms exposed to the effluent of problem drywall because no formal statistical sampling methods were used, and the sample size was small relative to the number of exposed smoke alarms. Additionally, only ionization smoke alarms were available for NIST to examine; and, therefore, conclusions about the effects of in-home contamination on other types of smoke alarm technologies are not possible.

The Set 1 and Set 2 smoke alarms were subject to unknown environments for approximately three years prior to testing at NIST. The indoor contaminant concentrations to which a particular smoke alarm may be exposed is a function of factors such as the nature and frequency of household cooking; indoor production of aerosols/particles from items such as paint, air fresheners, or pets; frequency of air changes within the home; local outdoor air quality; and the presence of smoking within the home. These factors are independent of the nature of the emissions from the drywall and might have affected the smoke alarms' performance.

The applicability of the Battelle Class IV corrosion accelerated aging protocol performed at Sandia National Laboratories is not evaluated here. However, the conditions listed in the prior paragraph are not part of the Battelle Class IV protocol, so the relationship between the Set 1/Set 2 smoke alarms and the relationship between the Set 3/Set 4 smoke alarms is not the same. Any such comparison is confounded further by differences between the models of smoke alarms available for testing. The FEDE tests conducted for the same units before and after accelerated aging (Set 3 and Set 4, respectively) examined only the isolated effect of the incremental chemical contaminants expected from the problem drywall over a period of 10 years.

The results cannot be extrapolated to all types of fire. There are many characteristics of fires that may affect smoke alarm performance, including smoke velocity, smoke particle size distribution and density, and smoke alarm orientation. In the tests described in this report, these parameters were held constant to the extent practicable in order to improve the comparability of the test results.

NIST did not assess the change in hazard to occupants or firefighters resulting from changes in smoke alarm performance.

UL 217 tests were conducted to address the first research question (Section III.A):

1. Do smoke alarms exposed to the effluent from problem drywall activate within allowable tolerances?

Of the 29 smoke alarms in Set 1, there were 15 of Model A, and seven each of Models B and C. Four units of each model were selected for testing in triplicate. The results of the 36 tests,[vii] shown graphically in Figure 7, demonstrate that all the smoke alarms activated well within the allowable MIC range in all tests.

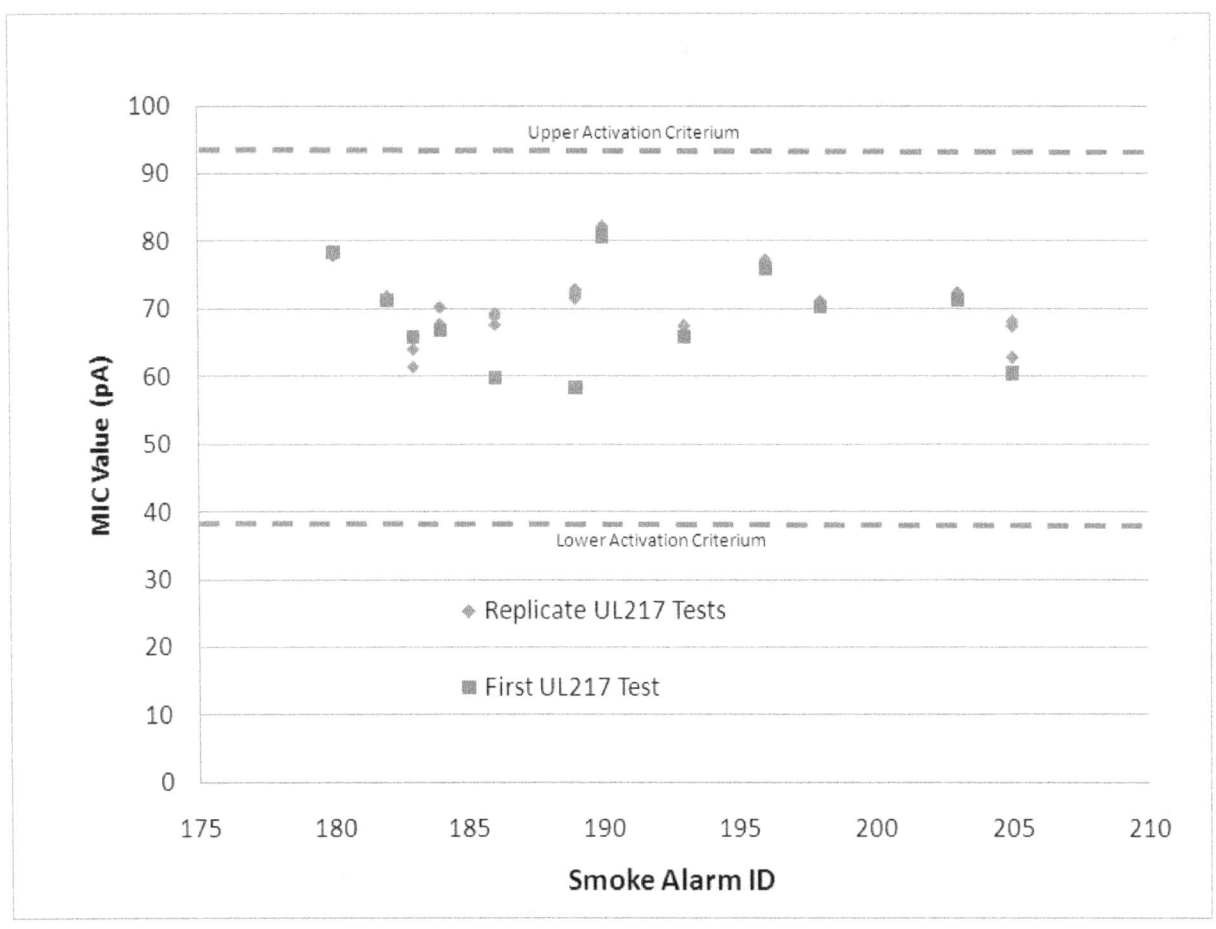

FIGURE 7: RESULTS OF UL 217 SMOKE ALARM TESTS

[vii] Some of the MIC values are identical, making it appear that there are fewer diamonds or squares.

C. FEDE TESTING

1. PURPOSE

FEDE tests were conducted to address the remaining two research questions (Section III.A):

2. Do smoke alarms exposed to the effluent from problem drywall perform differently from similar vintage smoke alarms from homes without problem drywall?

3. Does a 10-year exposure to a Battelle Class IV corrosivity environment, simulated through an accelerated aging test protocol, affect smoke alarm performance?

2. COMPARISON OF IONIZATION SMOKE ALARMS FROM HOMES

Figure 8 shows the aggregated results of the FEDE tests for all the Set 1 and Set 2 ionization smoke alarms. The values for the Set 1 smoke alarms might appear to fall into two groups that align with the more distinct grouping of the Set 2 values. Figure 9 sorts the data points[viii] by smoke alarm model and highlights the insight gained from an analysis based on the individual models of smoke alarms, rather than on the aggregated data for all models. The following text discusses findings from these test data. Supporting statistical detail is provided in Appendices A, B, C, and D.

Table 2 summarizes numerically the activation data for the Set 1 and Set 2 smoke alarms that were tested in the FEDE. As indicated in the last row of Table 2 and demonstrated in Figure D.7 in Appendix D, there was no statistically significant difference between the sensitivities of the full complement of smoke alarms in Set 1 and the full complement of smoke alarms in Set 2. However, the units of Model A in Set 1 were approximately 4 % more sensitive than units of Model A in Set 2 (+2.77 pA), while the units of Model B in Set 1 were approximately 8 % less sensitive than the units of Model B in Set 2 (-6.19 pA). Both conclusions were statistically significant at the 95 % confidence level. A similar comparison for Model C was not possible because there were none of these units in Set 2.

It is not possible to assign causality for the differences between the Set 1 and Set 2 ionization smoke alarms. The units of each model were not from the same manufacturer's batches in the two sets; the uniformity of the sensitivities of the alarms, when new, is unknown; and there was likely some variety in their in-home exposure conditions. Nonetheless, the difference in *direction* of the change in sensitivity between the Model A and Model B smoke alarms in Set 1 and Set 2, which came from different homes, suggests that the home environments were at least somewhat different.

Ionization	Set 1 (Mean/pA)	Set 1 (St Dev/pA)	Set 2 (Mean/pA)	Set 2 (St Dev/pA)	Difference (pA)	Difference (%)
Model A	74.64	4.50	71.87	1.92	2.77	3.71 %
Model B	74.67	4.18	80.86	2.47	-6.19	-8.29 %
Model C	78.93	4.28	-	-	N/A	N/A
All Set 1 & 2	75.79	4.72	76.27	5.04	0.48	0.63 %

TABLE 2: SUMMARY OF FEDE TEST DATA FOR SET 1 AND SET 2 SMOKE ALARMS

[viii] The tests were performed in triplicate. There appear to be fewer data points in both of these figures because many of the tests gave similar sensitivities, and the data points overlap.

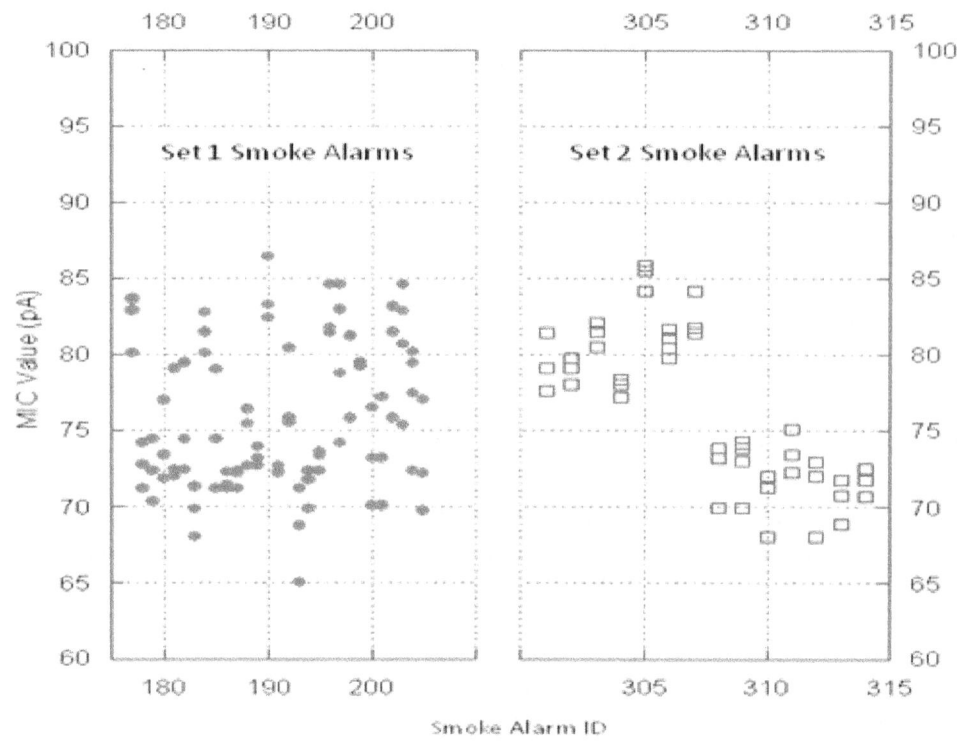

FIGURE 8: ALARM VALUES FOR ALL IONIZATION SMOKE ALARMS FROM HOMES
Left: Set 1; right: Set 2

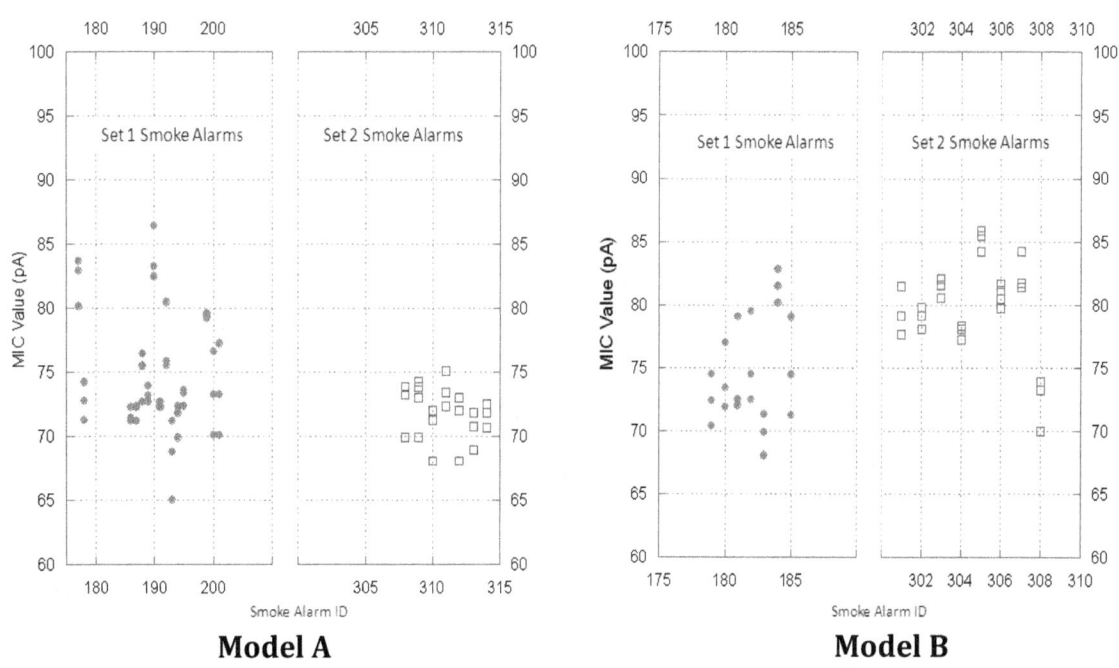

FIGURE 9: COMPARISON OF SET 1 VS. SET 2 ALARM VALUES MEASURED IN THE FEDE FOR MODELS A AND B
In each case, the Set 1 alarm data are represented by solid red circles and the Set 2 alarm data are
represented by open blue squares.

In addition to the changes in sensitivity for the same model alarms between Set 1 and Set 2, there were differences in the repeatability of the measurements. The standard deviations for the individual models in Set 2 were similar to each other and, as will be seen in Section VI.C, to the standard deviations for the Set 3 and Set 4 ionization smoke alarms. All of these standard deviations are comparable to the variation inherent in the FEDE, which arises from the stepwise nature of the sensitivity determination and the noise in the plateau values (Section IV.C). The environmental exposure history of these three Sets is quite different. Set 3 alarms were tested as purchased; the Set 4 units were exposed to a single, known, and invariant environment for a set period of time; and different units of the Set 1 alarms had been exposed to different home environments. The similarity of their standard deviations suggests that the environmental components that were not included in the Battelle Class IV exposure had a secondary effect on the alarms from the Set 1 homes.

By contrast, the variability in the Set 1 values is approximately double the values for the Set 2, Set 3, and Set 4 smoke alarms. This suggests that exposure differences within or among the Set 1 homes had significant variability that exceeded the variability in the fresh (Set 3) smoke alarms and the controlled Battelle Class IV exposure (Set 4). Formal interpretation of these values is limited by the small numbers of smoke alarm models, units of each model, and homes from which they came.

During the FEDE testing, NIST recorded two types of observations of smoke alarm performance that were unrelated to the sensitivity.

- Two of the 29 smoke alarms from Set 1 (No. 179 and No. 180) failed to generate an interconnect signal. Therefore, while the smoke alarms would sense the presence of smoke and produce an alarm sound themselves, connected units would not be notified and go into simultaneous alarm.

- Two other smoke alarms from Set 1 (No. 181 and No. 184) had AC power connection failures. Each device was able to be powered by its 9 V battery (the required backup power supply for residential smoke alarms), although the battery would likely be depleted at a faster rate in these smoke alarms than smoke alarms with a functional AC connection.

Neither type of failure is associated with the smoke-sensing mechanism. It was not possible to determine when these failures began. While it is possible that the in-home environment produced these four failures, the smoke alarms might have been manufactured incorrectly or impaired during installation or removal.

3. COMPARISON OF SMOKE ALARMS BEFORE AND AFTER ACCELERATED AGING TESTS

A. IONIZATION SMOKE ALARMS

Figure 10 shows the FEDE test results for the Set 3 and Set 4 ionization smoke alarms, Models D, E, and F. The data points are the mean MIC values at activation from triplicate tests for each of the six units of the three models of smoke alarms. Table 3 summarizes the activation values. Recall that the two Sets comprised the *same* smoke alarms, the only difference being that the Set 3 measurements were made when the alarms were new and the Set 4 measurements were made after exposure in the accelerated aging protocol. The following text discusses findings from these test data. Supporting statistical detail is provided in Appendices A, E, F, and G.

21

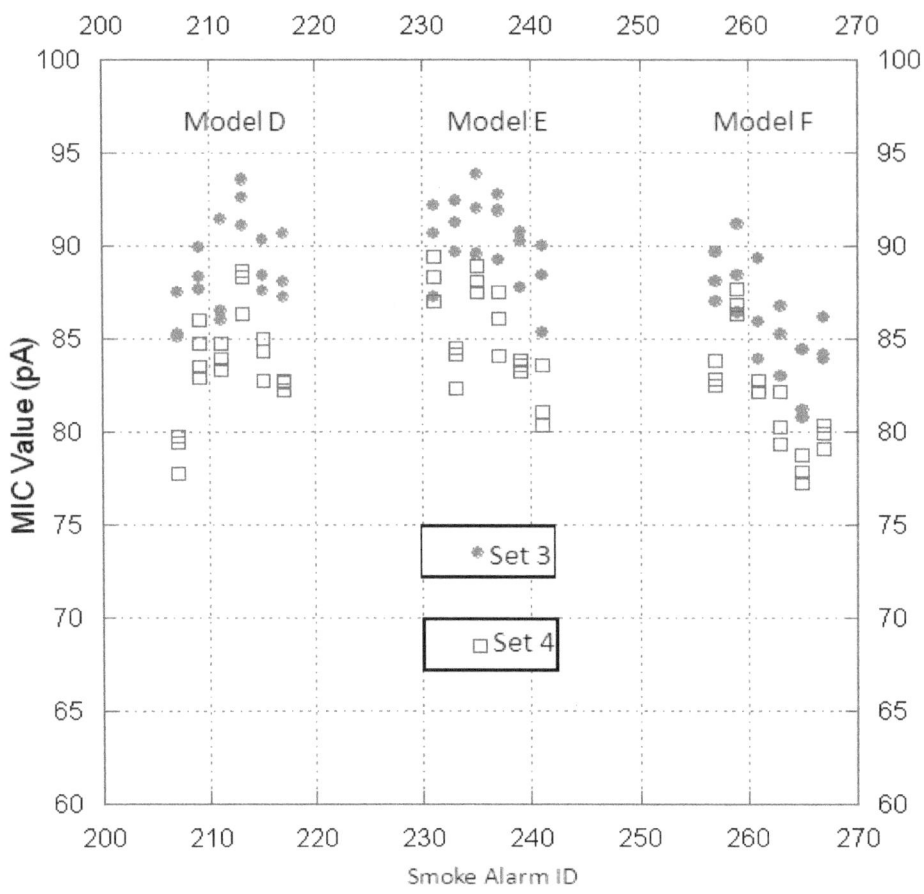

FIGURE10: COMPARISON OF IONIZATION SMOKE ALARMS MEASURED IN THE FEDE BEFORE AND AFTER ACCELERATED AGING TESTS

Ionization	Set 3 (Mean/pA)	Set 3 (St Dev/pA)	Set 4 (Mean/pA)	Set 4 (St Dev/pA)	Difference (pA)	Difference (%)
Model D	88.75	2.46	83.64	2.77	5.11	5.75 %
Model E	90.30	2.16	85.21	2.75	5.09	5.64 %
Model F	85.87	2.85	81.82	3.03	4.05	4.72 %
All Set 3 & 4 Ionization	88.31	3.08	83.56	3.12	4.75	5.38 %

TABLE 3: SUMMARY OF FEDE MIC TEST DATA FOR SET 3 AND SET 4 IONIZATION SMOKE ALARMS

All three models of ionization smoke alarms exhibited approximately 5 % decreases in sensitivity to smoke from the cotton wicks in the FEDE after exposure in the accelerated aging protocol. The differences of 4.05 pA to 5.09 pA were significant at the 95 % confidence level. The combined data for all three models of the Set 4 ionization smoke alarms were 4.75 pA less sensitive, a 5.4 % reduction in alarm sensitivity. As shown in Figure G.6 in Appendix G, this sensitivity change was also significant at the 95 % confidence level.

Although smoke alarm Models A and B (Set 1 and Set 2) are not the same as Models D, E, and F (Set 3 and Set 4), it is useful to note the similarity of the magnitude of the changes in sensitivity resulting from the different environmental exposures. All are in the nominal range of 3 pA to 6 pA, although in Set 1 the direction of change was different. Furthermore, while direct comparisons of activation currents between the FEDE protocol and the UL 217 protocol should consider the test differences discussed in Section V.D, a*s a matter of contextual comparison*, the reader should note that a 6 pA difference is numerically small compared to the allowable activation range for smoke alarms, which encompasses MIC values from 37.5 pA to 93 pA.

B. PHOTOELECTRIC SMOKE ALARMS

Figure 11 shows the FEDE test results for the Set 3 and Set 4 photoelectric smoke alarms, Models G, H, and I. The data points are the mean laser transmittance values at activation from triplicate tests for each of the six units of the three models. Table 4 summarizes the activation values. Recall that the two sets comprised the same smoke alarms, the only difference being that the Set 3 measurements were made when the alarms were new and the Set 4 measurements were made after exposure in the accelerated aging protocol. The following text discusses findings from these test data. Supporting statistical detail is provided in Appendices A, H, I, and J.

At first inspection, it appears that all three models were unaffected by the accelerated aging exposure. A rigorous statistical approach (Appendix J) showed that for the aggregated data for all three models and the individual data for Models G and H, the mean transmittance value at alarm was statistically indistinguishable at the 95 % confidence level for tests before and after exposure. For Model I, the accelerated aging exposure led to an increase in sensitivity that was less than 0.005 transmittance units. Although this difference is statistically significant at the 95 % confidence level, it is very small compared to the magnitude of the allowable performance range of 0.81 % to 0.98 % (Section V.C).

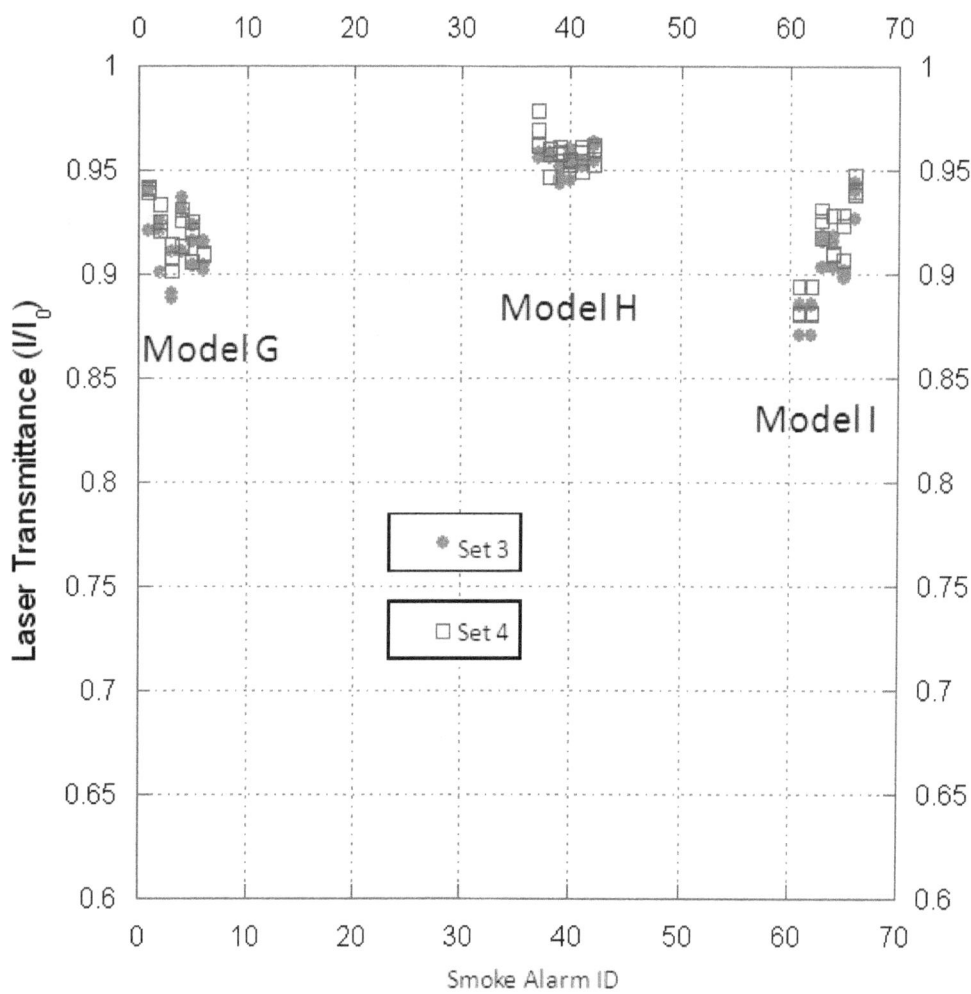

FIGURE11: COMPARISON OF PHOTOELECTRIC SMOKE ALARMS BEFORE AND AFTER ACCELERATED AGING TESTS

Photoelectric	Set 3 (Mean)	Set 3 (St Dev)	Set 4 (Mean)	Set 4 (St Dev)	Difference (trans.)	Difference (%)
Model G	0.916	0.016	0.921	0.013	-0.005	0.58%
Model H	0.954	0.006	0.959	0.007	-0.005	0.48%
Model I	0.904	0.021	0.912	0.022	-0.008	0.92%
All Set 3 & 4	0.926	0.027	0.931	0.025	-0.005	0.54%

TABLE 4: SUMMARY OF FEDE LASER TRANSMITTANCE TEST DATA FOR SET 3 AND SET 4 PHOTOELECTRIC SMOKE ALARMS

VI. CONCLUSIONS

This analysis provided answers to three questions:

1. Do smoke alarms exposed to the effluent from problem drywall activate within allowable tolerances?

   All of the Set 1 ionization fire smoke alarms, tested according to the protocol of UL 217, went into alarm within the permissible sensitivity limits. It is not possible to estimate whether these smoke alarms would continue to alarm within intended specifications if exposed to problem drywall effluent for a 10-year recommended lifetime.

   No photoelectric smoke alarms from homes were available for testing.

2. Do smoke alarms exposed to effluent from problem drywall perform differently from similar vintage smoke alarms from homes without problem drywall?

   Within the important limitations listed in Section VI.A, NIST found only minor (yet statistically significant) changes in the sensitivity of the two tested models of ionization smoke alarms between Set 1 and Set 2. The changes for the two models were in different directions, were not much larger than the measurement uncertainty in the sensitivity measurement, and were small compared to the range of permissible sensitivity. The household environments for the various smoke alarms are not known. Thus, it is not possible to ascribe the observed changes to any particular facet of the exposure history of the tested smoke alarms.

   Two of the 29 Set 1 smoke alarms sensed the presence of smoke, but would not send the signal to activate interconnected smoke alarms that might be located in other parts of the home. Two other Set 1 smoke alarms failed to operate under AC power, but operated properly under the required 9 V battery backup. There was insufficient information to determine when these failures began or whether they could be attributed to exposure to the emissions from problem drywall. Neither type of failure is associated with the smoke sensing mechanism.

3. Does a 10-year exposure to a Battelle Class IV corrosivity environment, simulated through an accelerated aging test protocol, affect smoke alarm performance?

   With the important limitations listed in Section VI.A, NIST measured a small, but statistically significant increase in the smoke concentration required to activate the tested ionization smoke alarms as a result of a two-week exposure to a Battelle Class IV corrosivity environment designed to simulate a 10-year exposure to the effluent from contaminated drywall. Two of the models of photoelectric smoke alarms tested in the NIST Fire Emulator/Detector Evaluator before and after exposure to the same environment and the same accelerated aging test protocol did not manifest statistically significant changes in the sensitivity to smoke from smoldering combustion. NIST measured a very small change in the performance of a third model. All of the changes in sensitivity were small compared to the allowable performance range.

VII. ACKNOWLEDGEMENTS

The authors gratefully acknowledge the contributions of several individuals. Tom Cleary (NIST) provided technical guidance on the FEDE operation and on smoke alarm performance. The experience of Michael Selepak (NIST) in running the FEDE apparatus was instrumental in the success of the project. Dr. Rob Sorenson (Sandia National Laboratories) led the accelerated aging tests (Set 4 smoke alarms) and provided (through CPSC) the description of the accelerated aging method contained in this report.

APPENDIX A: OVERVIEW OF STATISTICAL METHODS
AND FEDE DATA ANALYSIS

A. EXPLORATORY DATA EXAMINATION

For each set of smoke alarm sensitivity data from the FEDE, NIST performed an exploratory, qualitative data examination to see what factors influenced the responses and to identify any unusual data points and any other interesting features in the data. An exploratory data examination informs all subsequent statistical analyses and may influence the analyses that had been planned prior to the collection of the data if the assumptions underlying those analyses don't match the reality of the data obtained. The results of this stage of examination are expressed as impressions.

The factors examined in this stage included test date, position of the smoke alarm in the test apparatus, repeatability of replicate tests for each smoke alarm, uniformity of smoke alarms of the same manufacturer's model, and degree of similarity of different models of smoke alarms of the same operating principle. The figure of merit for the ionization smoke alarms is the MIC value at alarm. The figure of merit for the photoelectric smoke alarms is the laser transmittance at alarm.

The output from the exploratory examination of the FEDE data for the ionization smoke alarms from Set 1 and Set 2 are presented in Appendices B and C, respectively. The examination of the output from the smoke alarms from Set 3 and Set 4, the purchased smoke alarms tested when new and after accelerated aging, are discussed in Appendices E and F (ionization smoke alarms) and Appendices H and I (photoelectric smoke alarms), respectively.

B. QUANTITATIVE ANALYSIS

The second phase of the analysis is a numerical quantification of differences in smoke alarm performance in the FEDE. NIST used two types of statistical methods to quantify these differences. For the comparisons of the average performance of multiple smoke alarms within one model, or across models, a Monte Carlo resampling method called the nonparametric bootstrap [10] was used. In contrast, standard two-sample confidence intervals (CI) based on Student's t distribution [11] were used for comparisons of the performance of individual smoke alarms from Set 3 and Set 4, before and after accelerated aging.

Bootstrap methods were chosen for the comparisons that averaged across multiple smoke alarms because of the statistical complexities of the data collection process. These complexities include the fact that there is correlation between some of the individual smoke alarm results, the mixture of continuous and discrete elements in determining smoke alarm response in each test, and the unequal numbers of smoke alarms and tests in some of the sets of data.

The correlation between some pairs of test results arises because two smoke alarms are measured in each run and the measured values are actually read from the test apparatus itself, based on when each unit alarms, rather than coming directly from the smoke alarms themselves. This means that the readings from two smoke alarms can be perfectly correlated if they alarm (even at different times) within the same smoke plateau.

Bootstrap methods handle these complexities through the use of an empirical probability distribution to describe the behavior of each estimator of interest. This empirical distribution is constructed directly from the observed data, capturing the performance of the actual measurement process, rather than relying on the assumption that the estimator will follow some known distribution such as the normal, Poisson, or Weibull distributions.

The trade-off for such specificity in the probability distributions used to determine the statistical significance of smoke alarm performance is the need for relatively large data sets from which to construct the distributions. Bootstrap distributions are constructed by creating thousands of "new" samples from the one sample observed by randomly (re)sampling from the observed data with replacement. The estimator of interest is then computed for each bootstrap sample making its approximate distribution to be obtained directly from the values taken on by the ensemble of bootstrap estimates.

Confidence intervals for the parameter estimated by the chosen statistic can then be computed using various methods, such as cutting off the upper and lower tails of the distribution to achieve the desired level of confidence (the percentile method). For example, to compute a 95 % confidence interval, it is typical to cut off the extremes of the distribution at the upper and lower points corresponding to 2.5 % of the probability in each tail.

For the bootstrap analyses presented here, however, a different method known as the "bootstrap-t" was used to compute confidence intervals for each difference in smoke alarm performance. In the bootstrap-t method, the bootstrap estimators are first centered on the value of the estimator of interest obtained from the original sample and scaled by the standard deviation of the bootstrap distribution. Then upper and lower quantiles, $t_{\alpha/2}^{B}$ and $t_{1-\alpha/2}^{B}$, of this standardized distribution are determined to achieve the desired confidence level, $100(1-\alpha)$ %, where $0<\alpha<1$ is the analyst's specification of the probability that the confidence interval will not contain the true value of the measurand (e.g., $\alpha = 0.05$ for a confidence level of 95 %).

Finally, the confidence interval is formed by computing the confidence bounds $\left(y - t_{\alpha/2}^{B} s_y, \ y + t_{1-\alpha/2}^{B} s_y \right)$, where y is the estimator from the originally observed sample and s_y is the standard deviation of the bootstrap estimators. The bootstrap-t was used because it often more accurately attains the desired confidence level than the percentile method does.

Bootstrap confidence intervals for the various differences in smoke alarm performance of interest are given in Appendices D (Set 1 and Set 2), G (Set 3 and Set 4 for ionization smoke alarms), and J (Set 3 and Set 4 for photoelectric smoke alarms). For the data from smoke alarm Set 1 and Set 2, the difference of bootstrap means for smoke alarms from homes with problem drywall and homes with domestic drywall were computed, and the distribution of their difference was used to obtain the final confidence intervals. For the data from smoke alarm Set 3 and Set 4, where matched pairs of smoke alarms before and after accelerated aging were available, the bootstrap confidence intervals were constructed based on individual smoke alarm differences.

Confidence intervals for each smoke alarm in Set 3 and Set 4 were computed using the standard Student's t distribution confidence interval because the assumptions underlying such an analysis are likely to be met and there was not enough data on each smoke alarm to use bootstrap methods. Confidence intervals for individual differences in smoke alarm performance are presented in Appendices G and J.

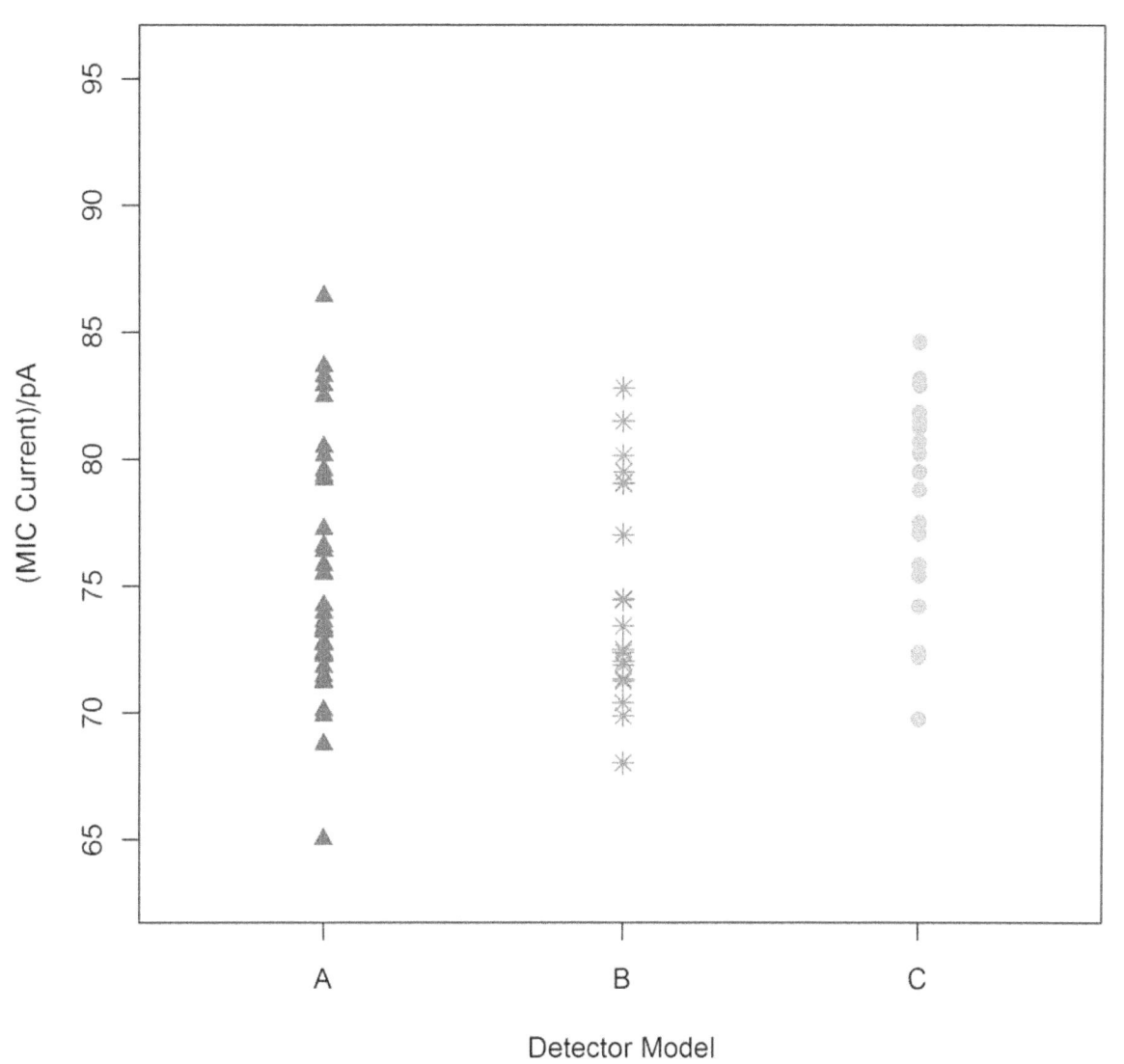

FIGURE B.1: SMOKE ALARM RESPONSES VS. MODEL FOR SET 1 IONIZATION SMOKE ALARMS

The responses from multiple tests of the 29 smoke alarms are color coded by model. From visual inspection of these data, the responses of all three models are very similar with respect to both their average responses and their levels of random variation between smoke alarms and measurements.

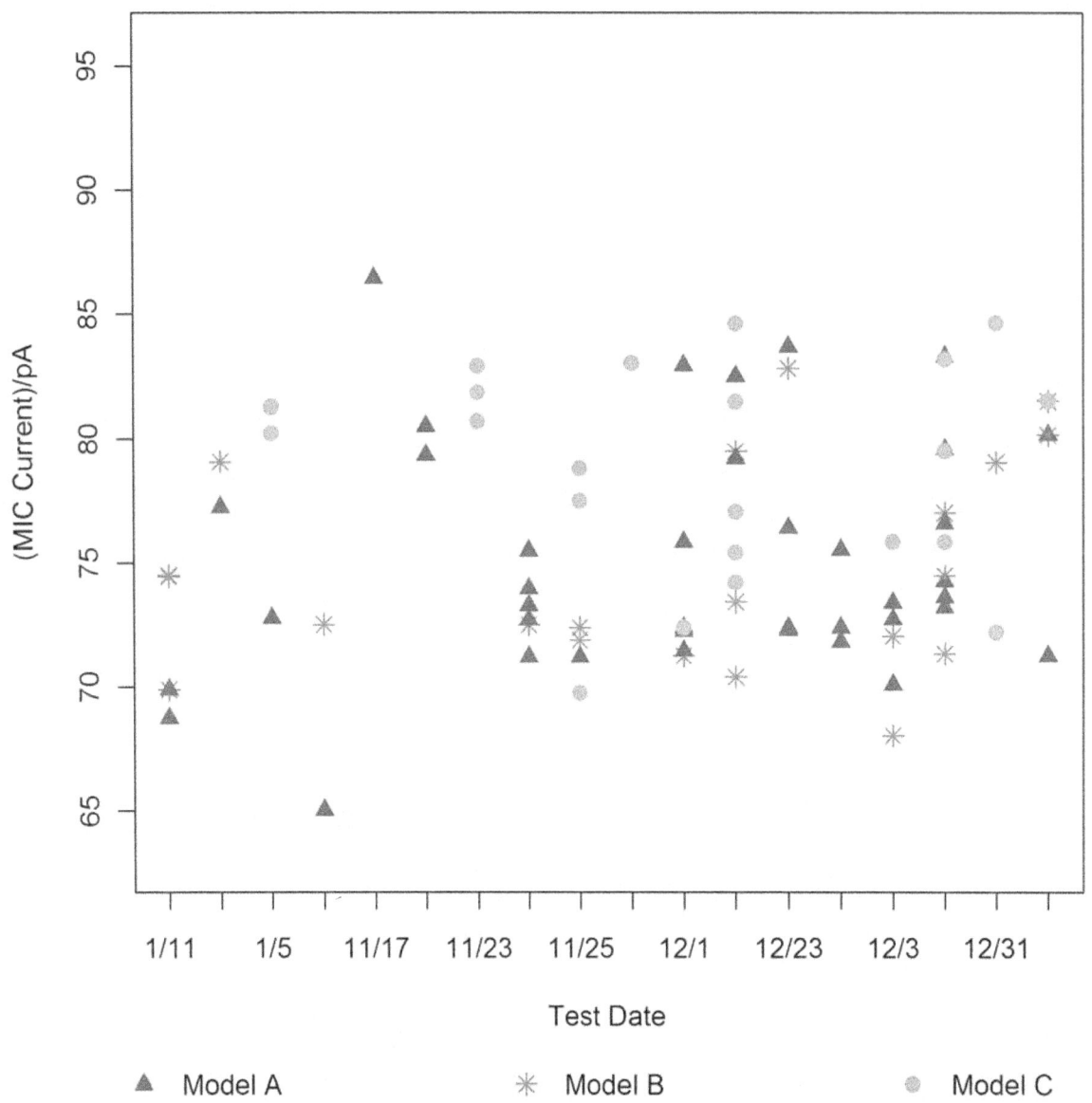

FIGURE B.2: PLOT OF SET 1 SMOKE ALARM RESPONSES VS. TEST DATE

Color coding indicates the smoke alarm model. The tests on all dates look similar with respect to both mean response and random variation, indicating that no significant random variation from day-to-day impacted the data.

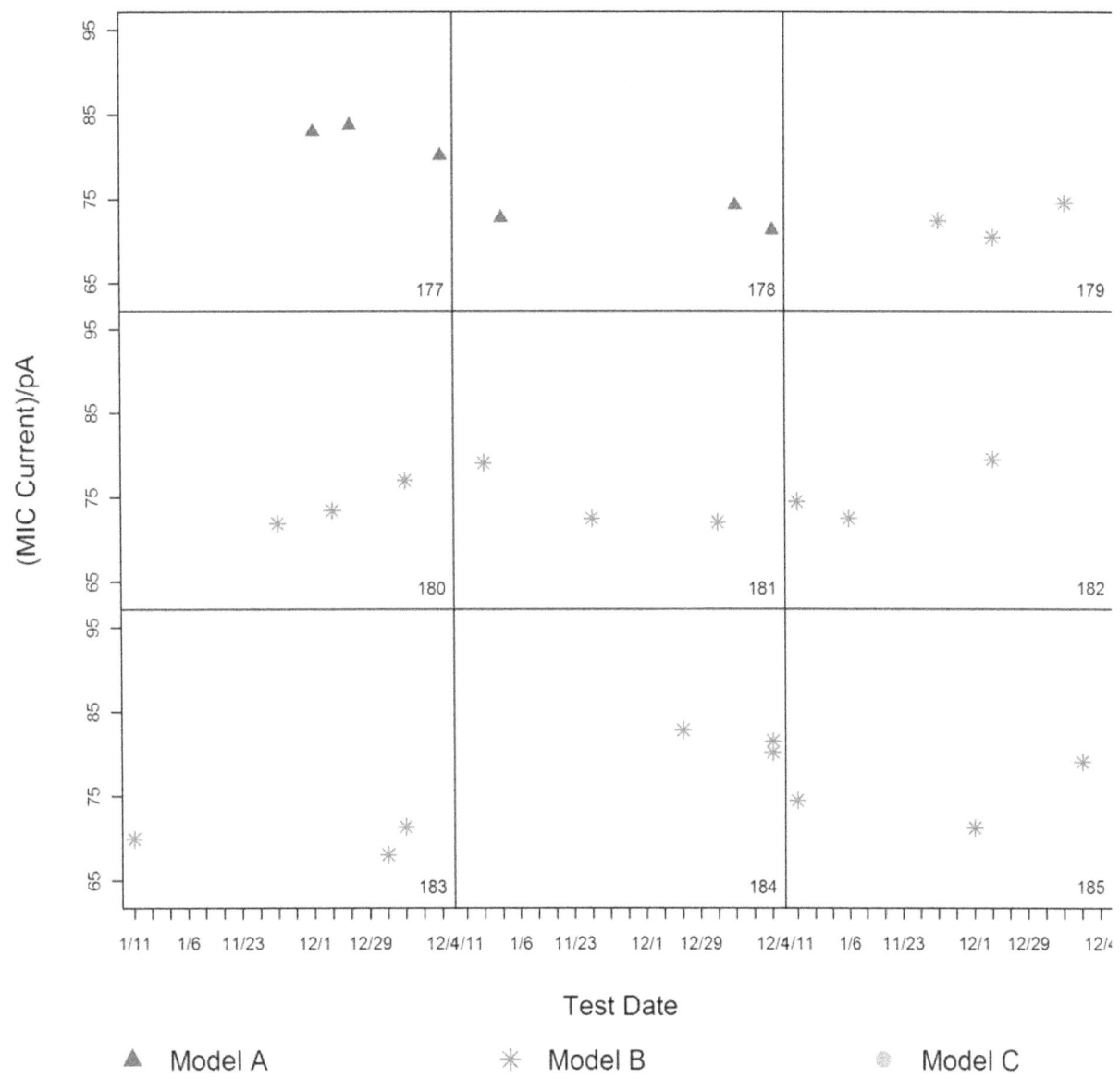

FIGURE B.3: SET 1 SMOKE ALARM RESPONSE VS. TEST DATE BY INDIVIDUAL UNIT

Unit identification numbers are given in the lower right corner of each plot. These plots show in more detail how each unit responded to tests performed on different dates. Looking across the plots for each model, the responses on different dates look relatively consistent.

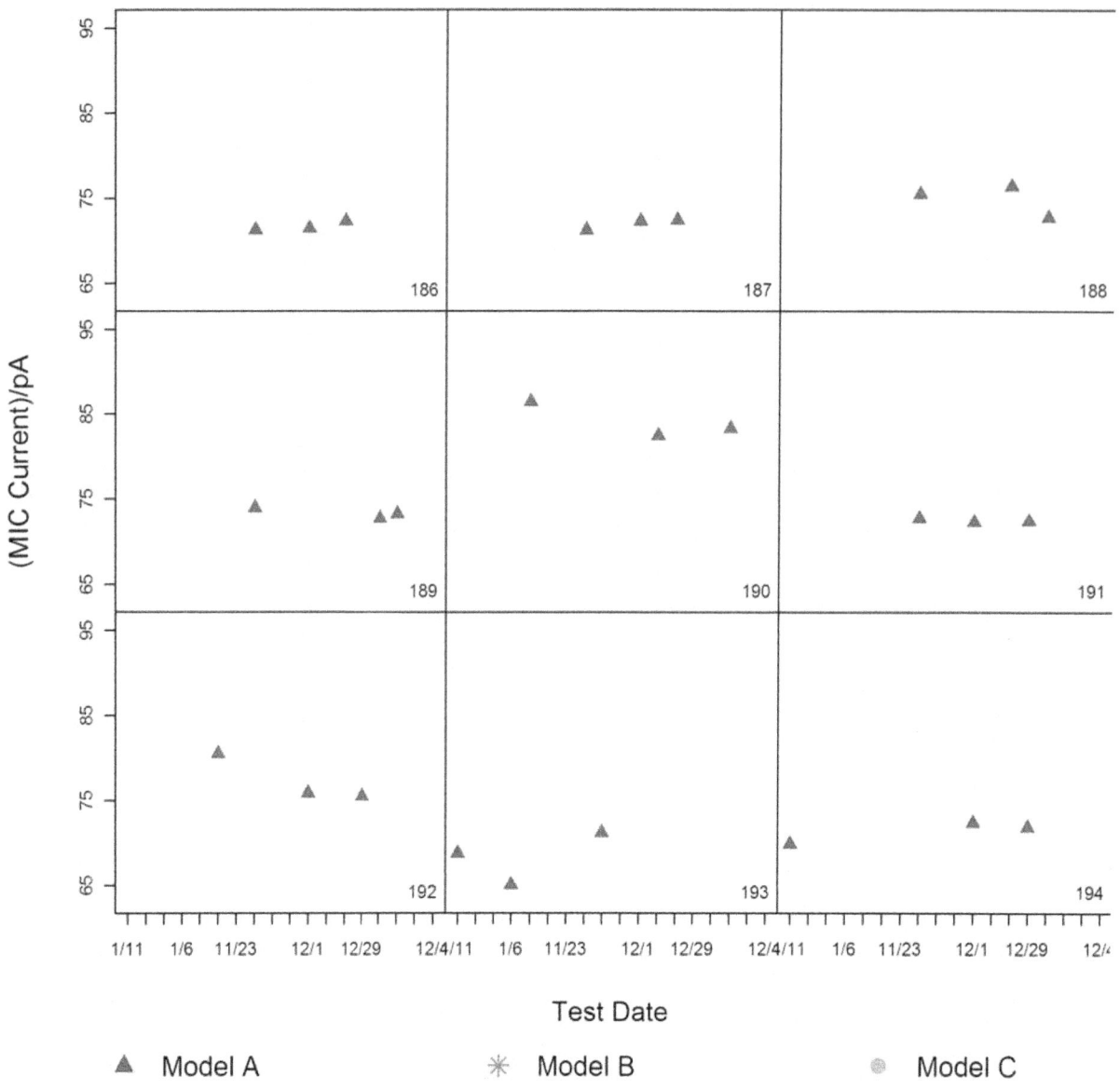

FIGURE B.3 CONTINUED: CAPTION AND INTERPRETATION ON PREVIOUS PAGE

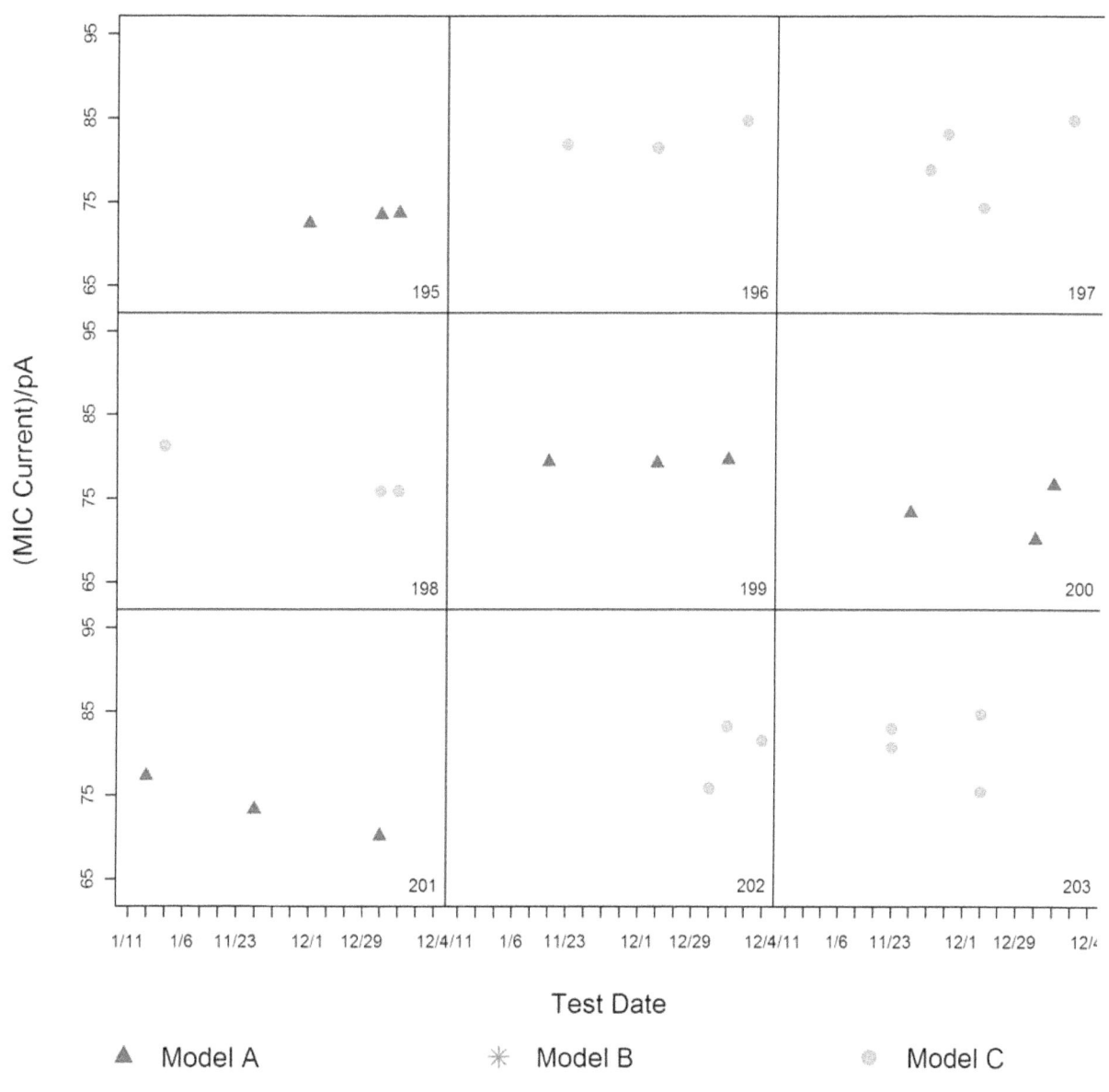

FIGURE B.3 CONTINUED: CAPTION AND INTERPRETATION ON PREVIOUS PAGE

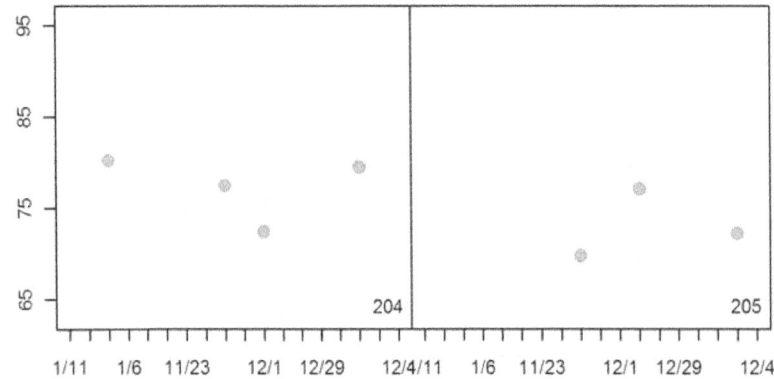

Test Date

▲ Model A   ＊ Model B   ● Model C

FIGURE B.3 CONTINUED: CAPTION AND INTERPRETATION ON PREVIOUS PAGE

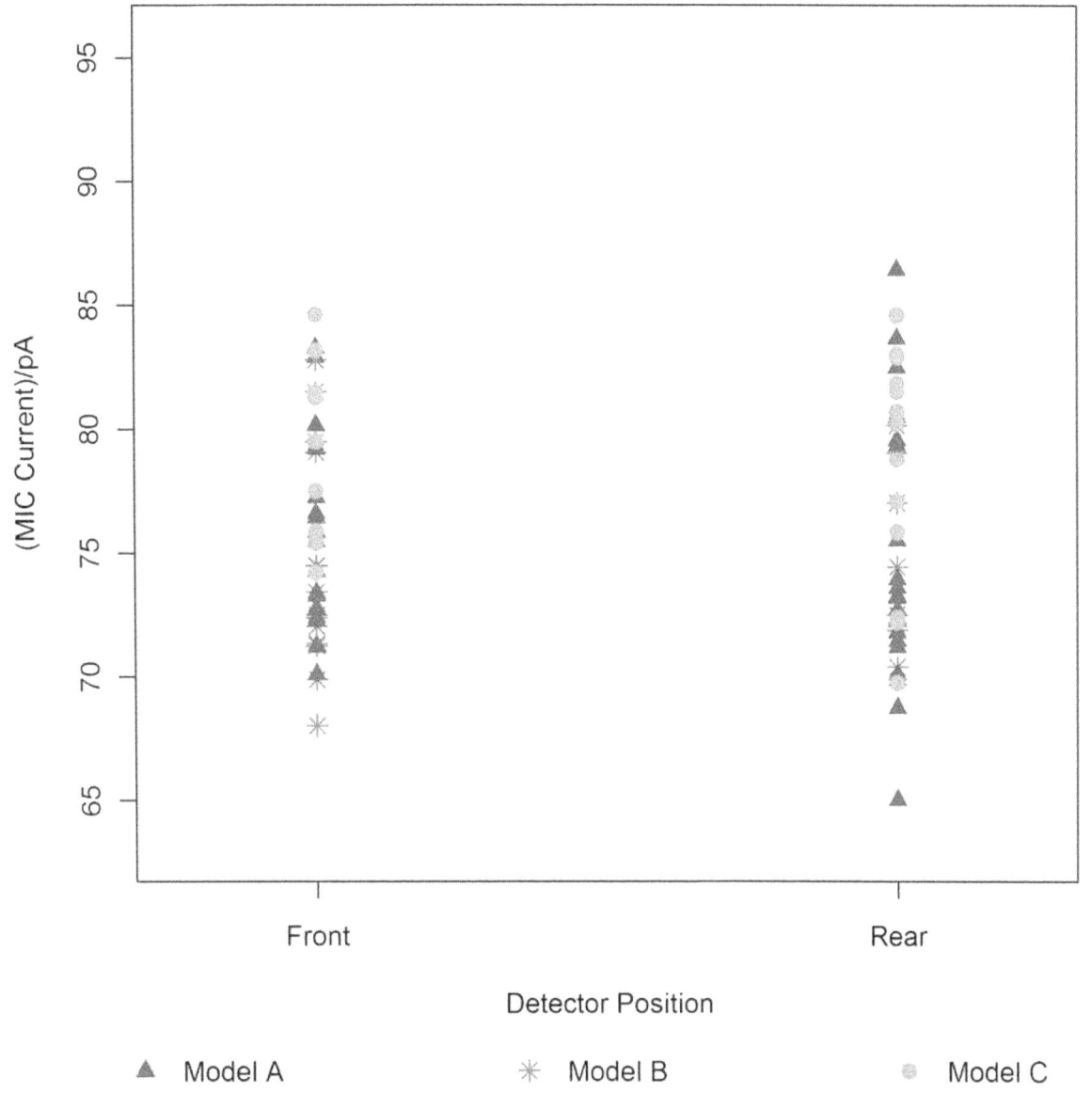

FIGURE B.4: COMBINED SET 1 SMOKE ALARM RESPONSES VS. POSITION IN THE FEDE

The smoke alarm model is indicated by the color coding. The similar responses for each position indicate that this factor does not affect the results on average for any of the three smoke alarm models.

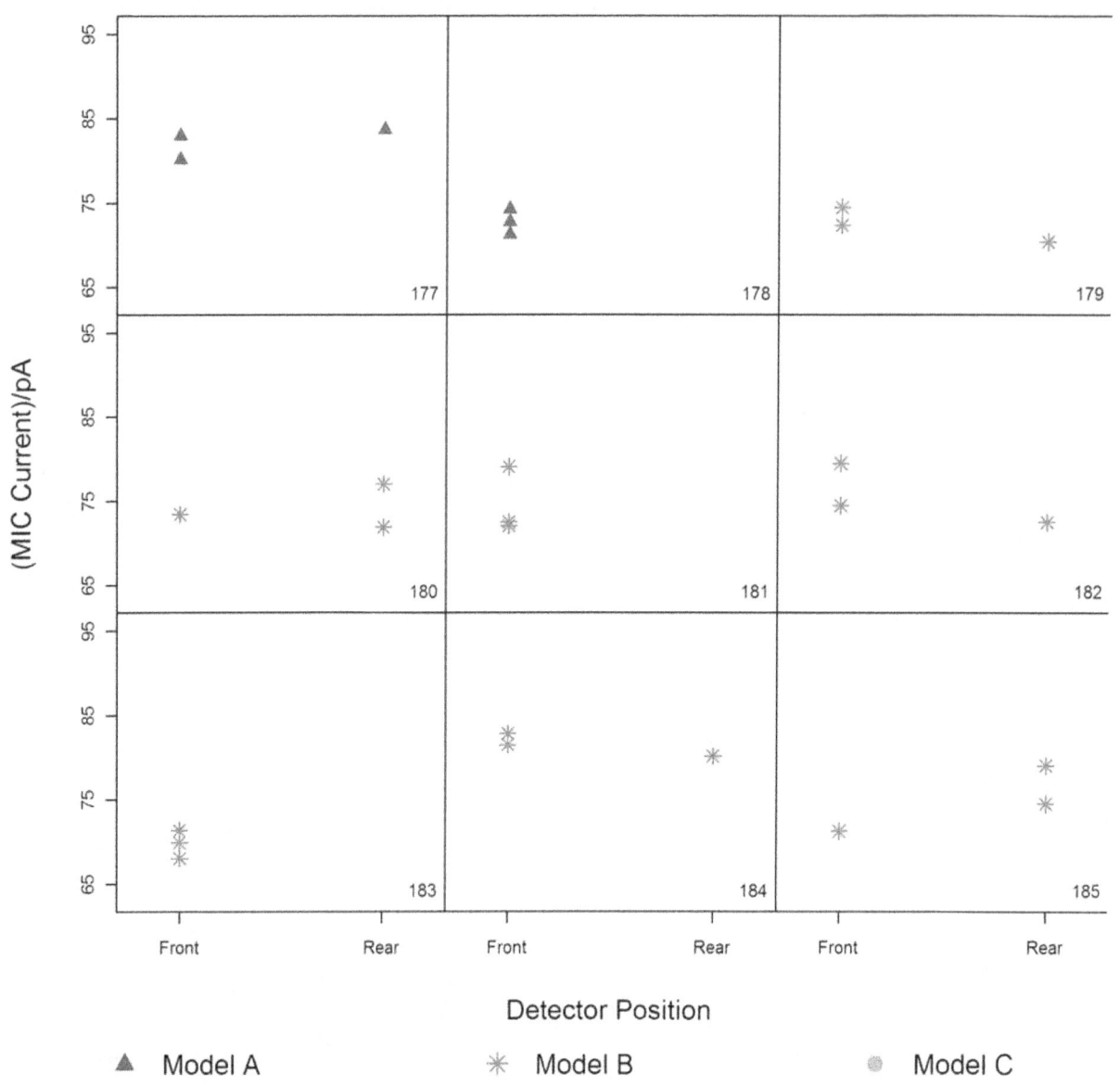

FIGURE B.5: INDIVIDUAL SET 1 SMOKE ALARM RESPONSES VS. POSITION IN THE FEDE

The smoke alarm models are indicated by the color coding, and the smoke alarm identification numbers are shown in the lower right corner of each plot. These plots provide more detail about individual smoke alarm responses to position in the FEDE test apparatus. In all cases, the response for each smoke alarm looks relatively consistent, regardless of position.

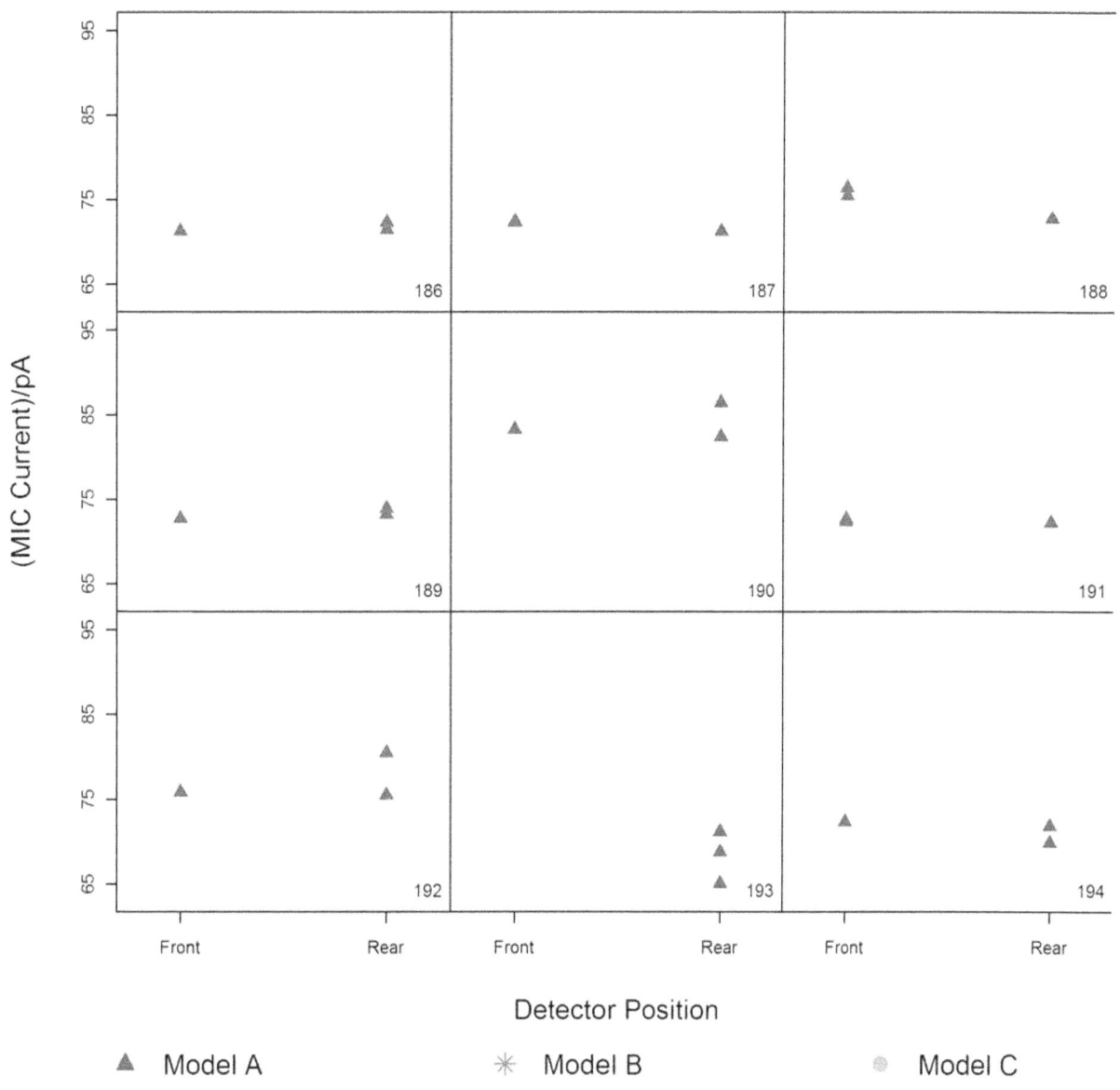

FIGURE B.5 CONTINUED: CAPTION AND INTERPRETATION ON PREVIOUS PAGE

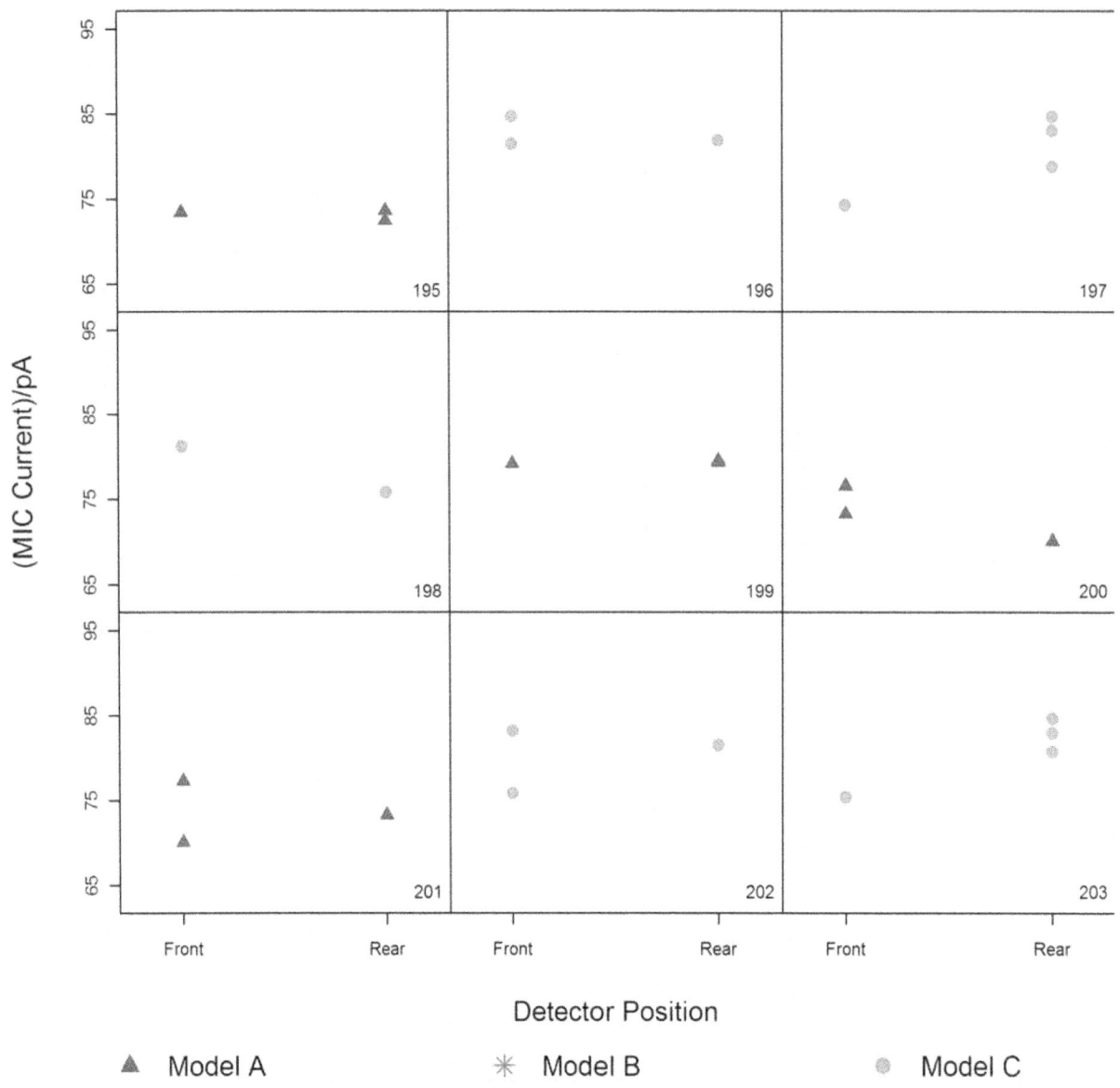

FIGURE B.5 CONTINUED: CAPTION AND INTERPRETATION ON PREVIOUS PAGE

38

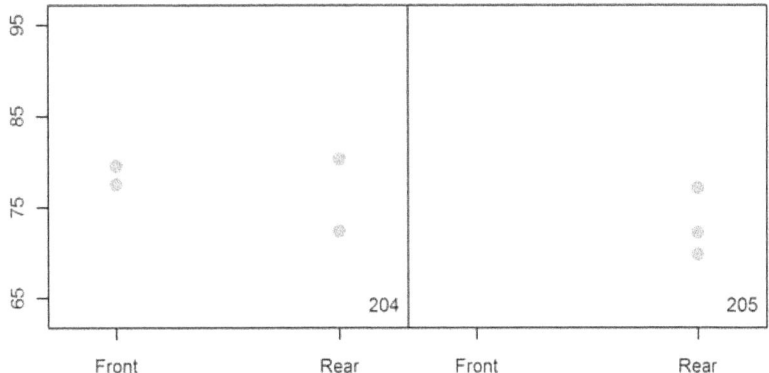

(MIC Current)/pA

Detector Position

▲ Model A   ＊ Model B   ◉ Model C

FIGURE B.5 CONTINUED: CAPTION AND INTERPRETATION ON PREVIOUS PAGE

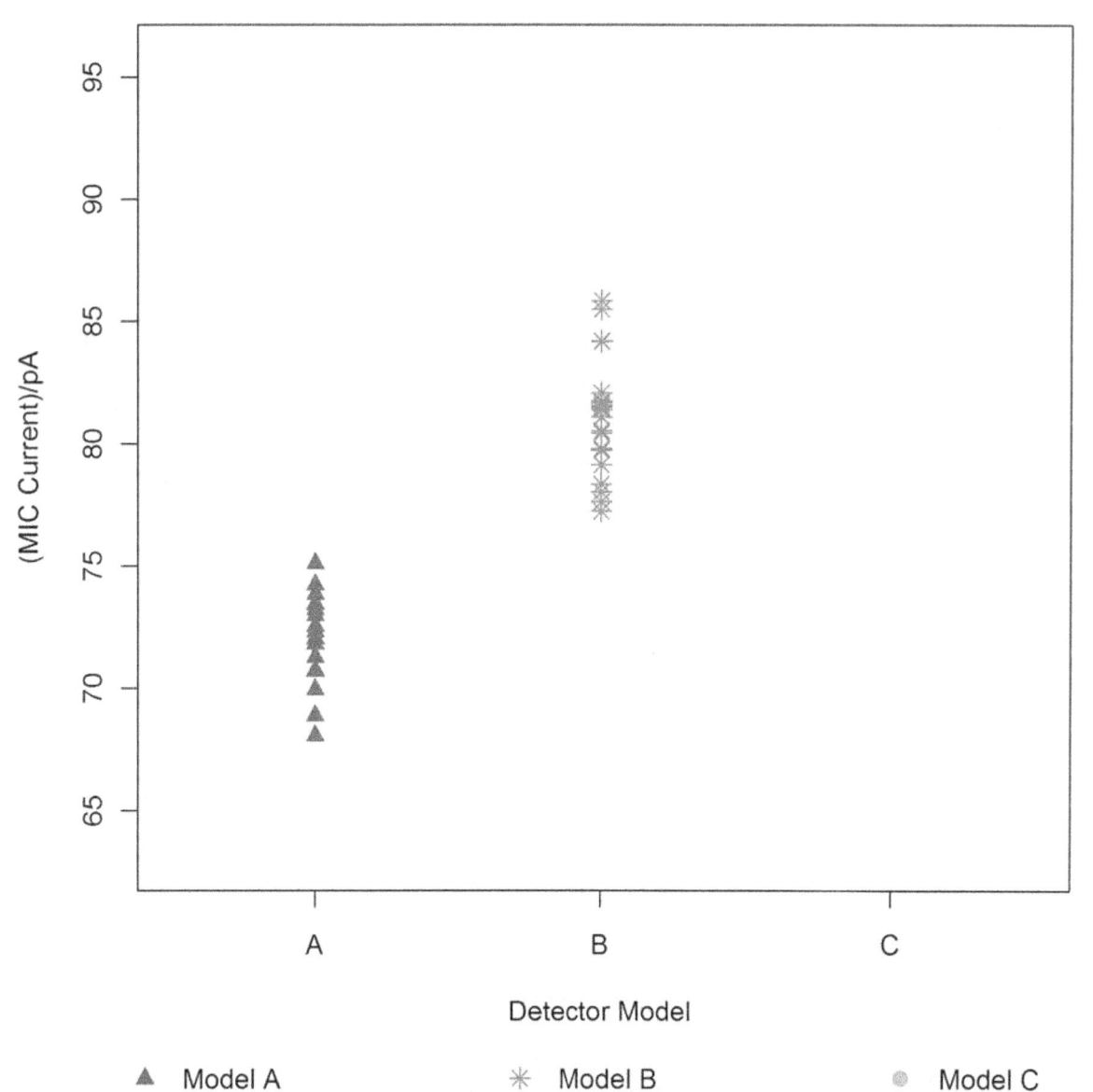

FIGURE C.1: SET 2 SMOKE ALARM RESPONSES VS. MODEL

The responses from multiple runs on 14 different smoke alarms are color coded by model. Compared to the Set 1 results from Figure B.1, these smoke alarms show a distinct difference in response between models. The random scatter in these smoke alarms appears to be lower than the scatter seen in the Set 1 units as well. Comparing individual units, however, the random variation looks the same for both sets of smoke alarms. This suggests that unit-to-unit variation may be greater in the Set 1 smoke alarms, perhaps due to the larger number of units in Set 1.

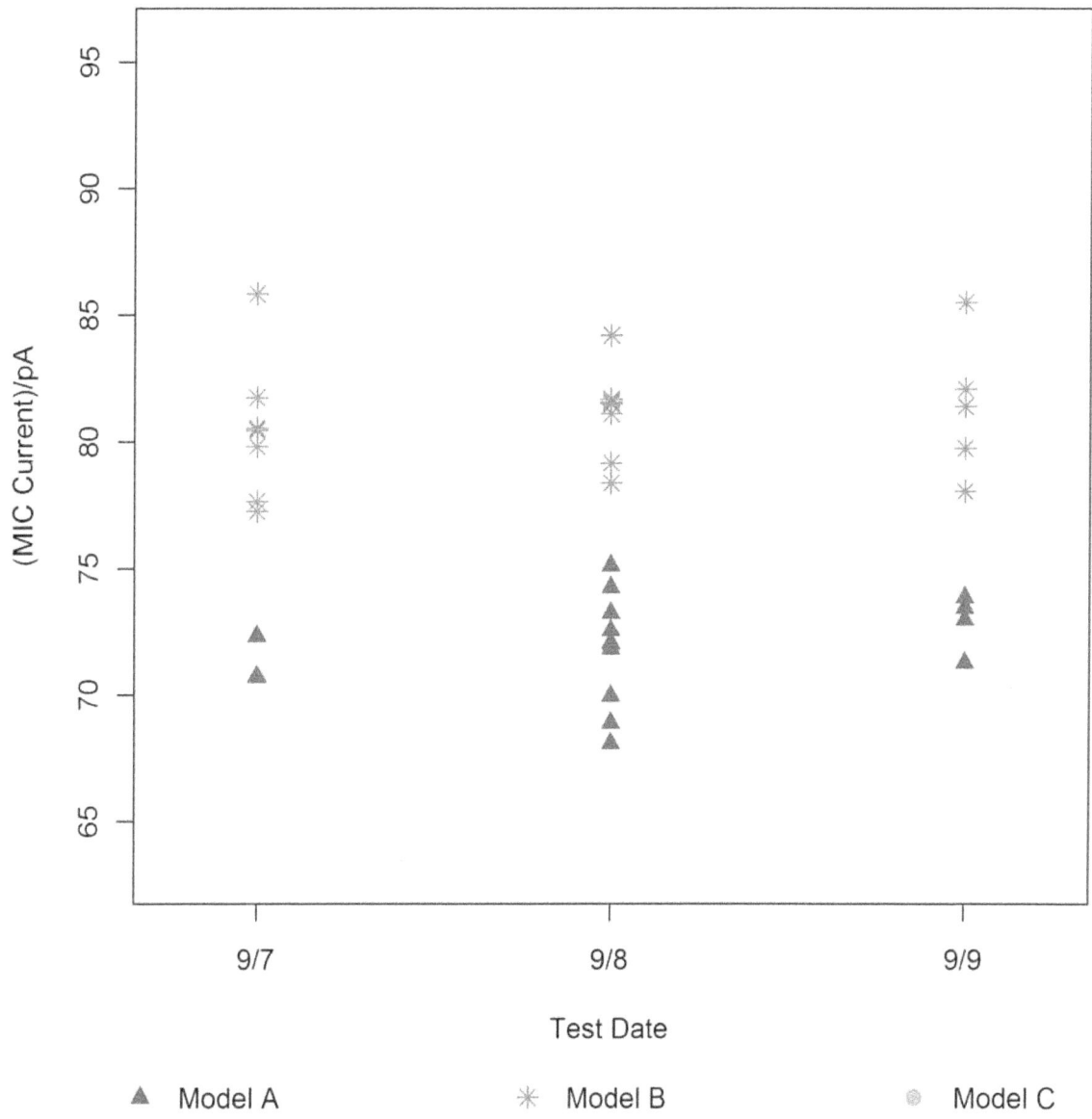

FIGURE C.2: SET 2 SMOKE ALARM RESPONSES VS. TEST DATE

The color coding indicates the smoke alarm model. The measurements made on different days look consistent across days for all three models. The difference in responses from the two models is seen clearly here as well.

41

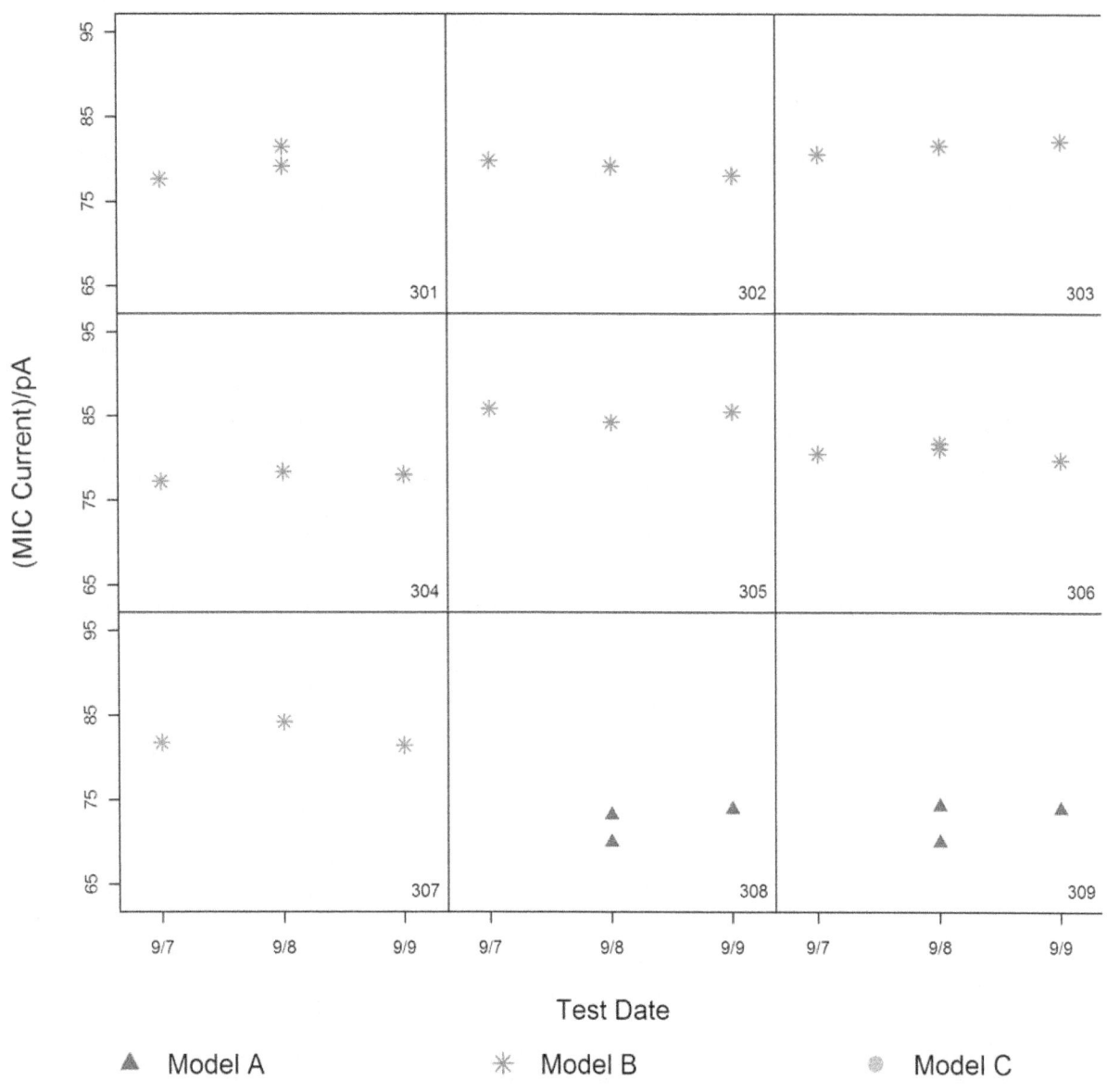

FIGURE C.3: SET 2 SMOKE ALARM RESPONSE VS. TEST DATE BY UNIT

The smoke alarm models are indicated by color. The unit identification numbers are given in the lower right corner of each plot. In this case, each unit was tested on only one day (and in one position in the FEDE). However, looking across the plots for each model, the responses on different days look quite consistent overall.

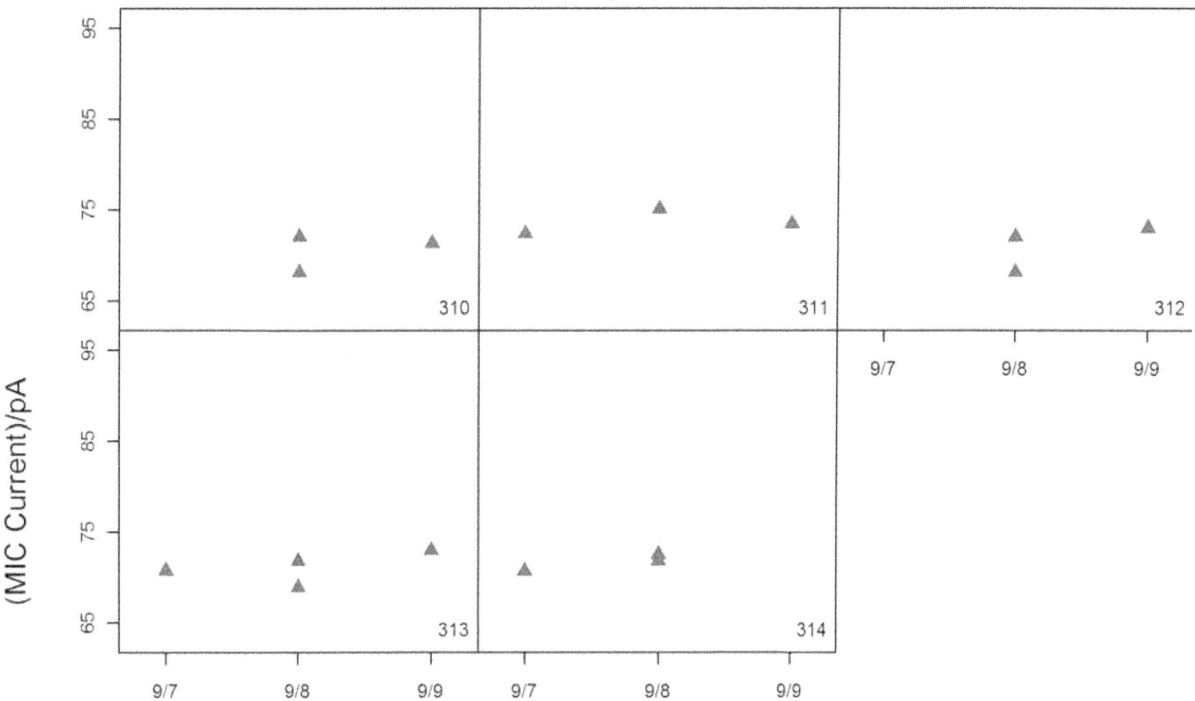

(MIC Current)/pA

Test Date

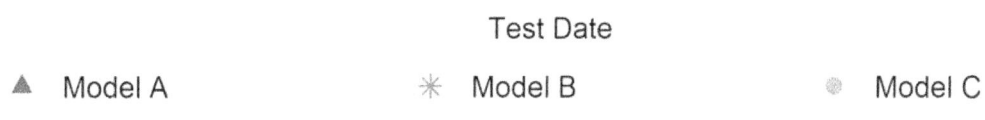

FIGURE C.3 CONTINUED: CAPTION AND INTERPRETATION ON PREVIOUS PAGE.

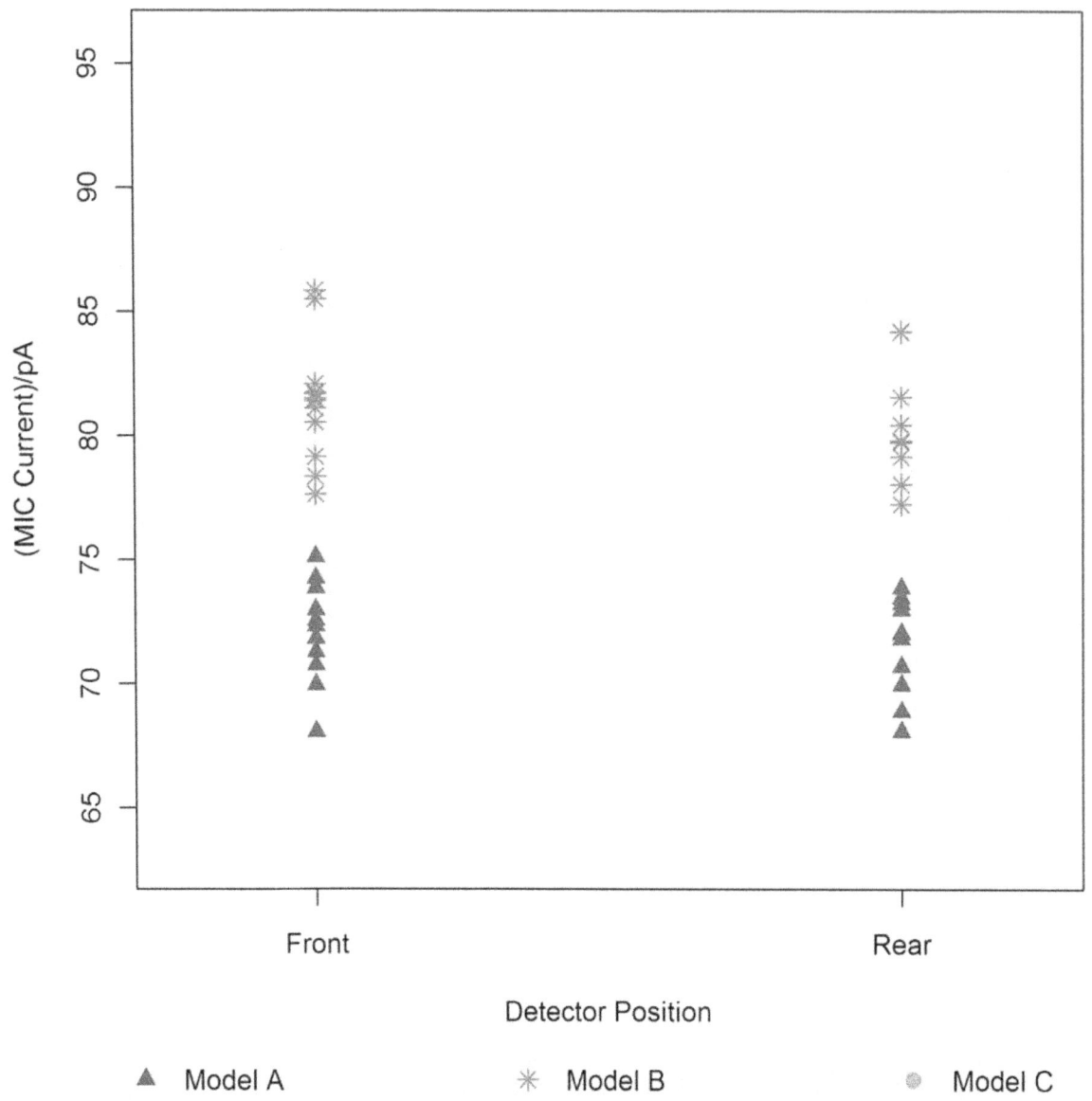

FIGURE C. 4: SET 2 SMOKE ALARM RESPONSE VS. POSITION IN THE FEDE

The smoke alarm model is indicated by the color coding. The similar responses for each position indicate that this factor does not affect the results on average for either of the two models.

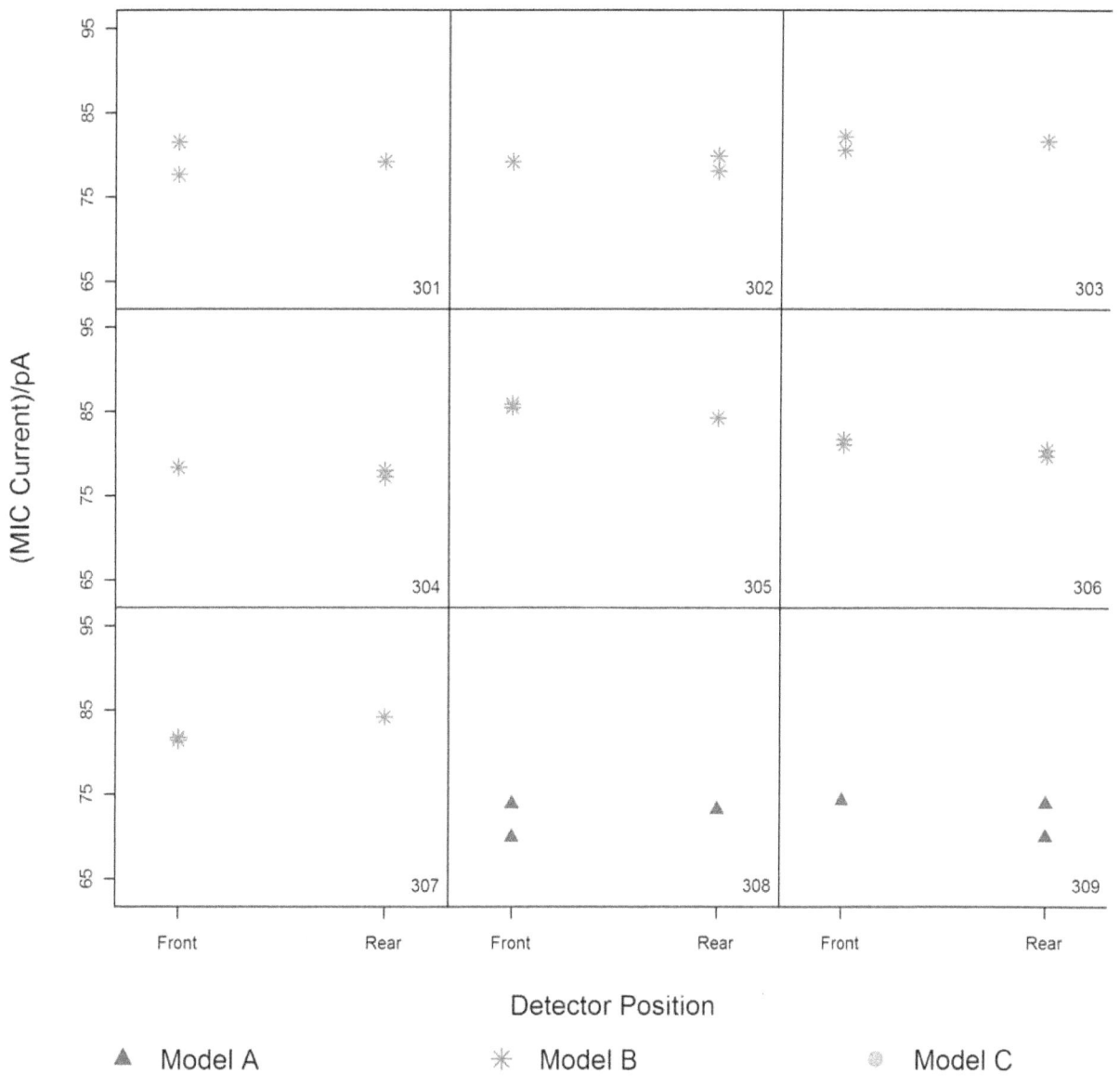

FIGURE C. 5: SET 2 SMOKE ALARM RESPONSES VS. POSITION

The smoke alarm models are indicated by the color coding and the individual unit identification numbers are shown in the lower right corner of each plot. These plots provide more detail about individual unit responses. In all cases, however, the responses across smoke alarms look consistent for each model of smoke alarm, regardless of position.

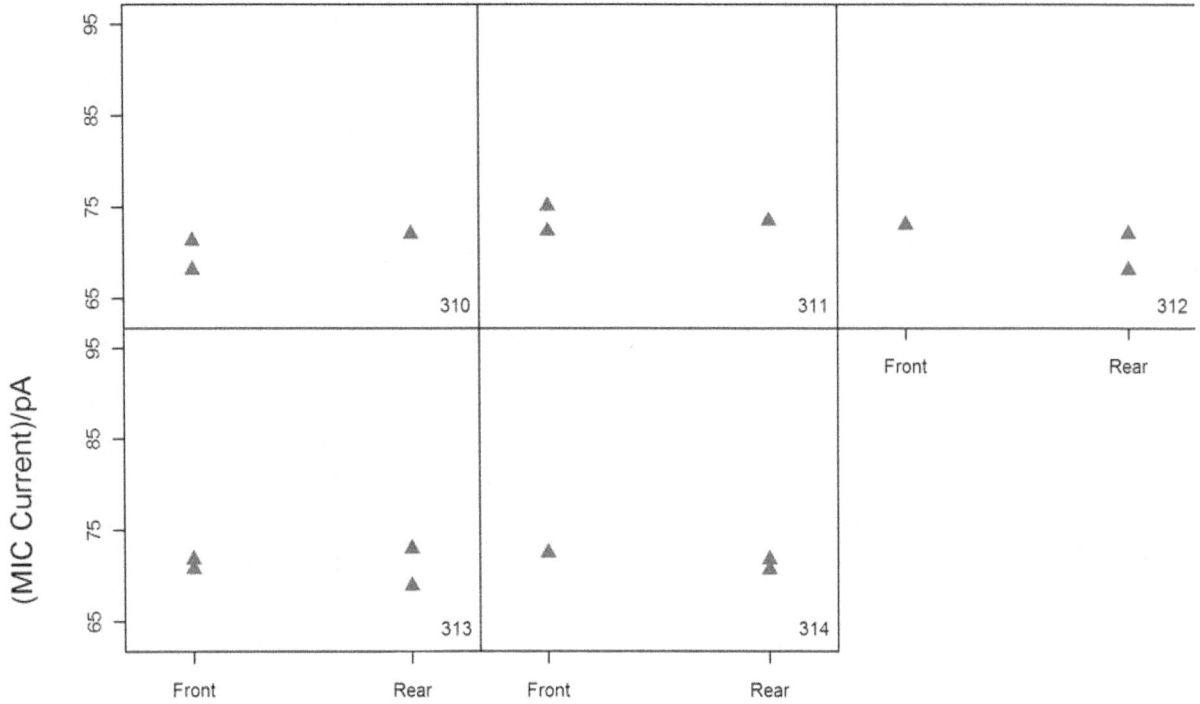

Detector Position

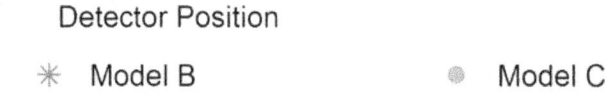

FIGURE C.5 CONTINUED: CAPTION AND INTERPRETATION ON PREVIOUS PAGE

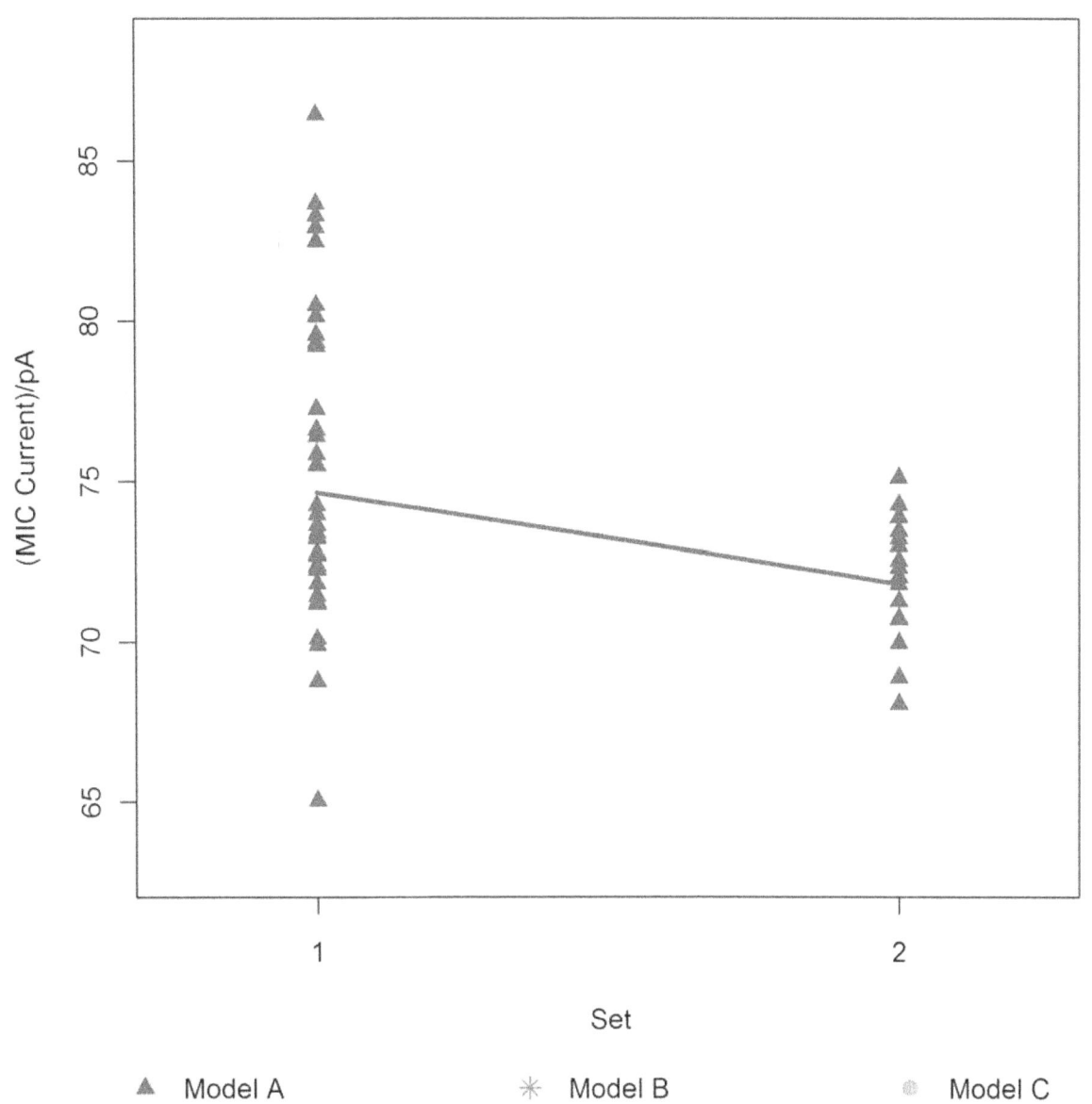

FIGURE D. 1: SMOKE ALARM RESPONSES FOR SET 1 AND SET 2 (MODEL A)

Responses from multiple tests of Model A smoke alarms are shown. The diagonal line between the two sets of data connects the mean responses. The line gives an indication of the average effect of the two groupings of exposures. Although some units responded similarly, the smoke alarms in Set 1 have a larger range of responses and a higher mean, leading to an increase in MIC for Set 1 smoke alarms relative to Set 2 smoke alarms on average.

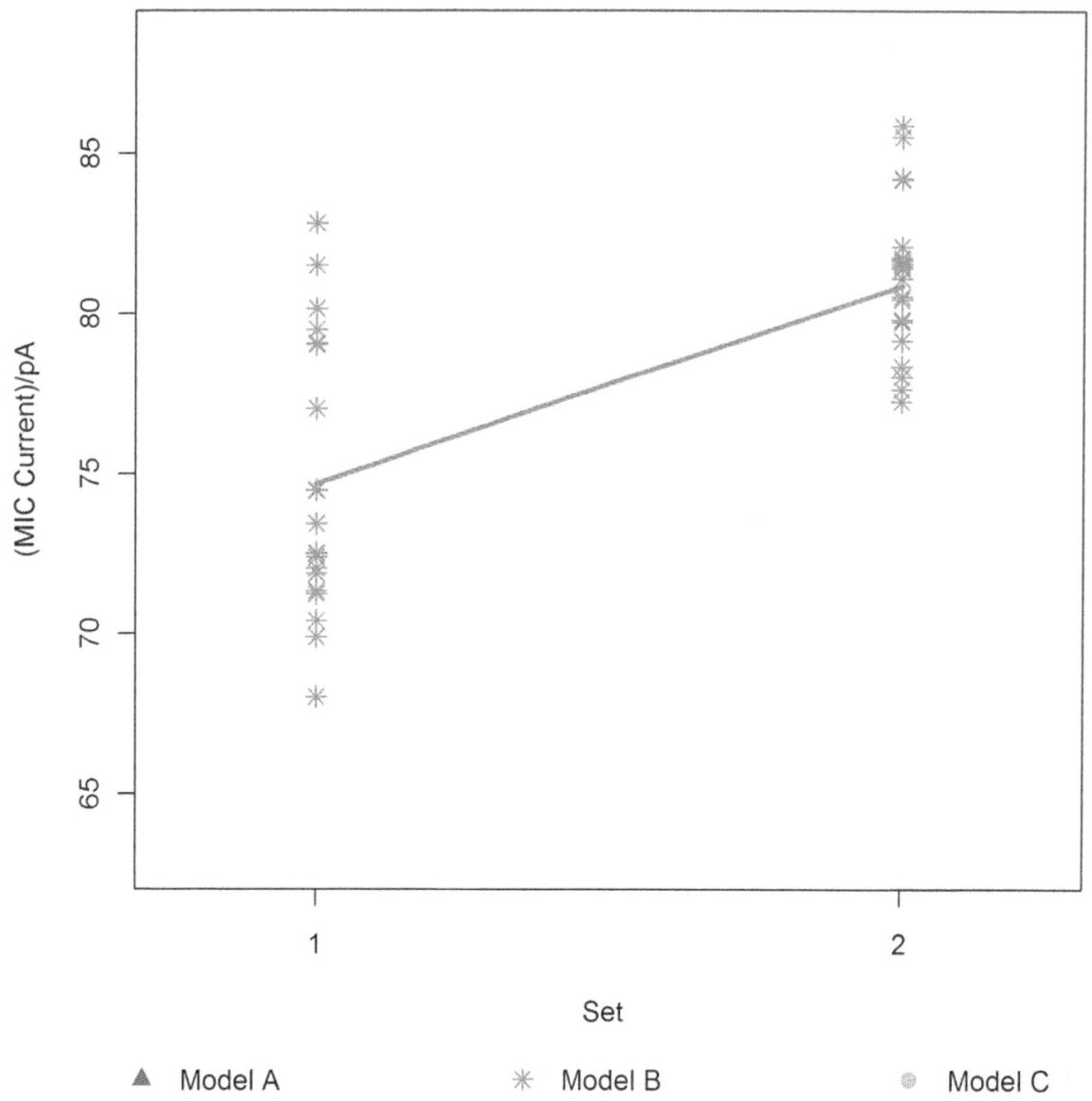

▲ Model A ✳ Model B ◉ Model C

FIGURE D. 2: SMOKE ALARM RESPONSES FOR SET 1 AND SET 2 (MODEL B)

Responses from multiple tests of Model B smoke alarms are shown. The diagonal line between the two sets of data connects the mean responses. The line gives an indication of the average effect of the two groupings of exposures. In constrast to the results for Model A shown in Figure D.1, the Set 1 Model B units tend to have lower MIC responses.

Ths slopes of the lines in Figures D.1 and D.2 are opposite in sign, suggesting substantial differences between the two sets of smoke alarms. This could result from an initial difference in sensitivity between different batches of the same model, differences in the overall residence environments to which the different units were exposed, and/or differences in reaction to different types of installed drywall. Due to the absence of information regarding both the pristine smoke alarms and the history within the homes, it is not possible to resolve the cause of the difference in slopes.

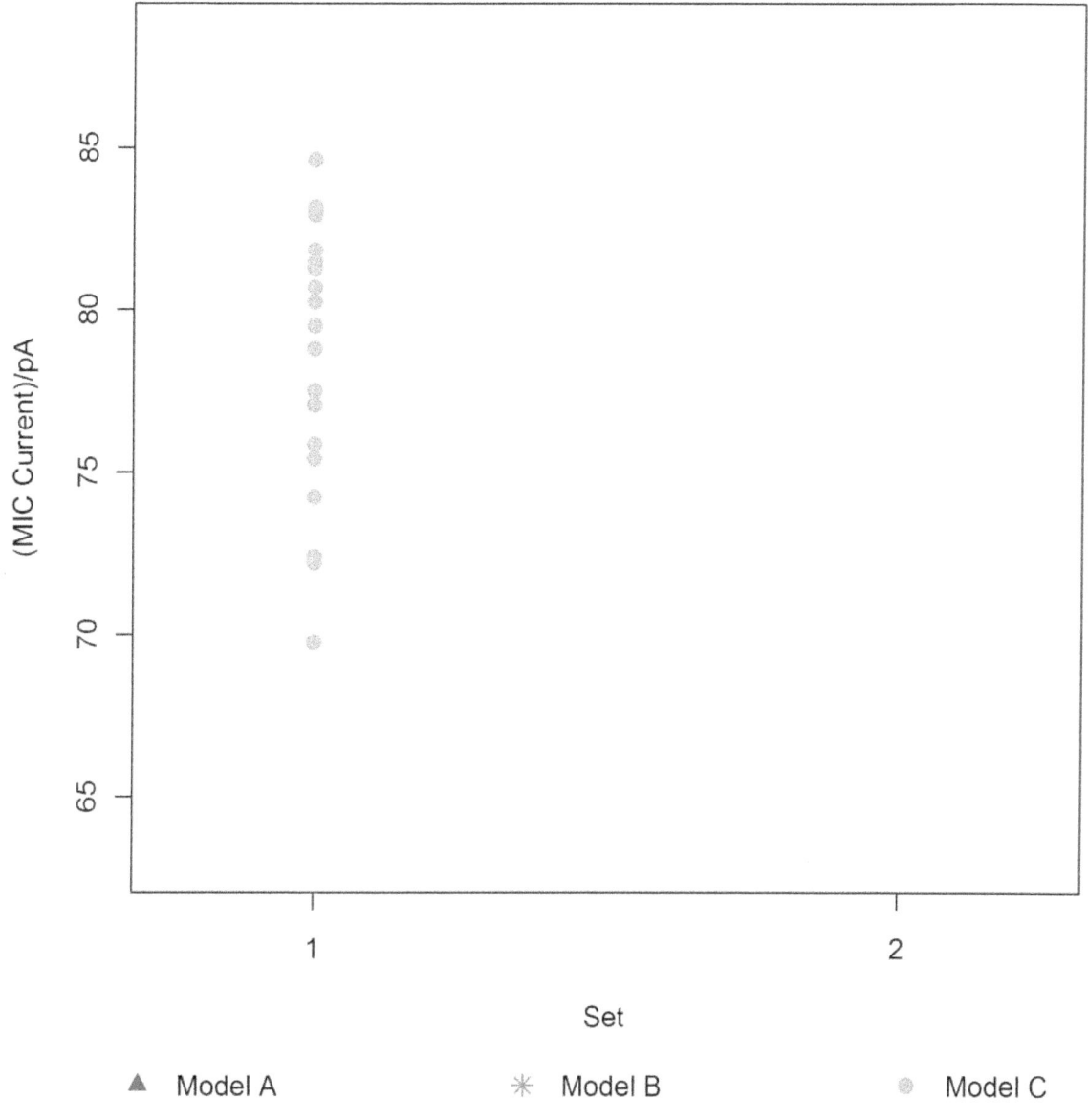

FIGURE D. 3: SMOKE ALARM RESPONSES FOR SET 1 (MODEL C)

The responses from multiple runs on Model C smoke alarms are shown here. Because there were no Model C units in Set 2, no model-specific conclusions about the performance of these smoke alarms can be investigated.

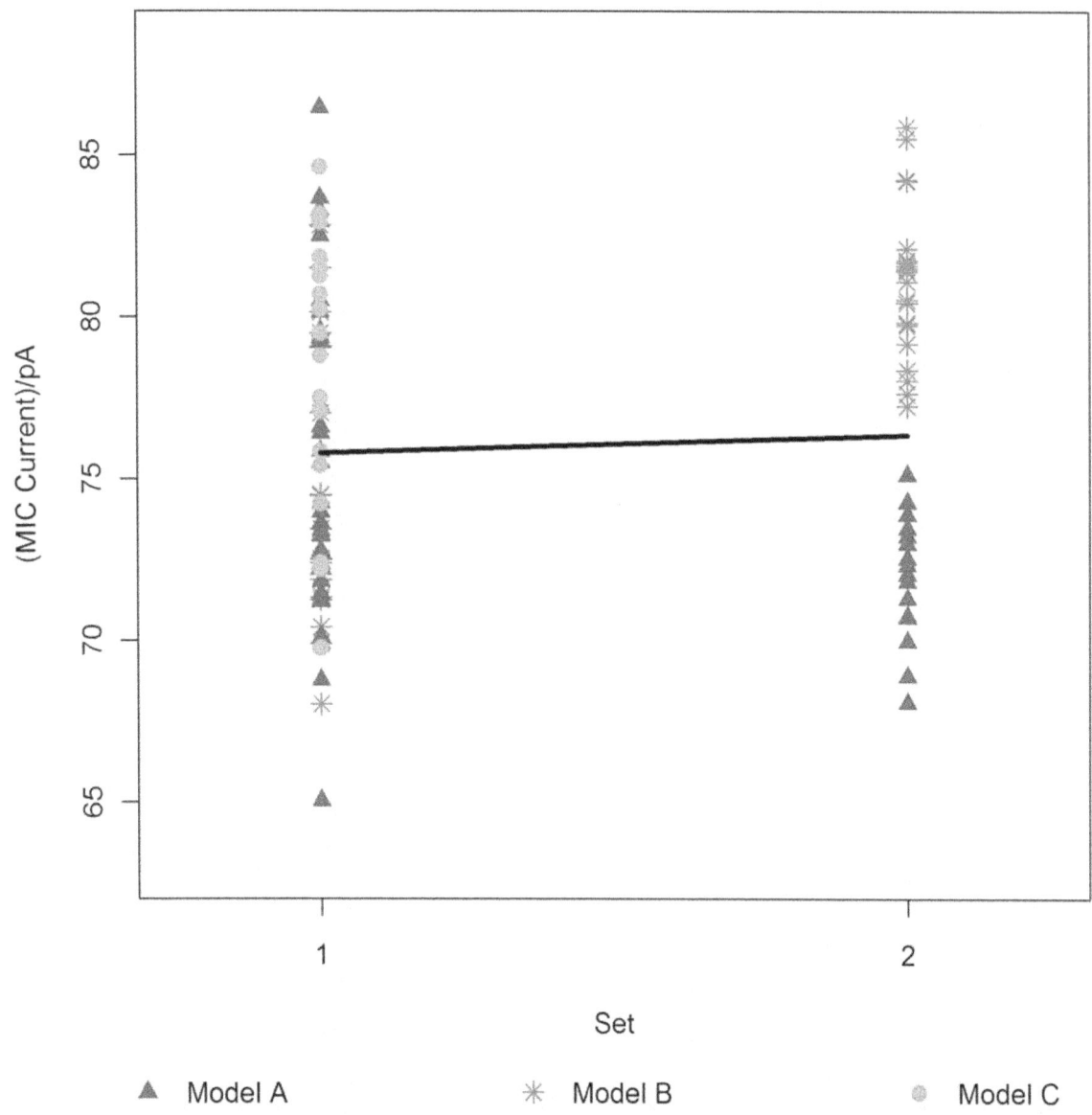

FIGURE D. 4: SMOKE ALARM RESPONSES FOR SET 1 AND SET 2 (ALL MODELS)

The responses from tests of all smoke alarms are shown. The line between the two sets of data connects the mean responses. The line gives an indication of the average effect of the two groupings of exposures. Looking at all models together, the increase in the MIC seen for Model A smoke alarms and the decrease seen for Model B smoke alarms essentially cancel, yielding no change on average. The fact that the Model C unit responses are at about the same level as Models A and B means inclusion of Model C data does not shift the overall mean in either direction.

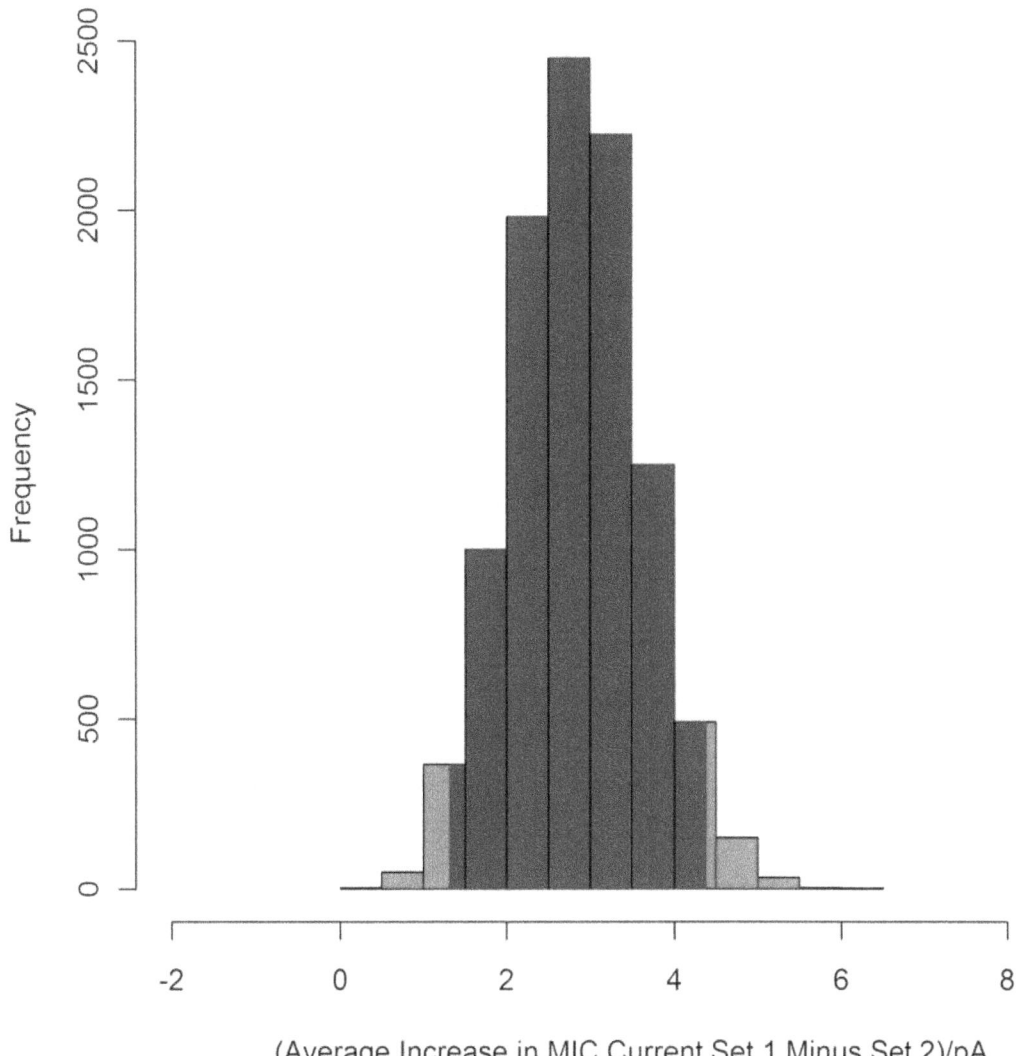

(Average Increase in MIC Current Set 1 Minus Set 2)/pA

95 % Confidence Interval for Average MIC Current Increase: (1.3 , 4.4) pA

FIGURE D. 5: HISTOGRAM SHOWING THE DISTRIBUTION OF RESAMPLED ESTIMATES OF THE AVERAGE
DIFFERENCE IN SMOKE ALARM RESPONSE FOR ALL MODEL A SMOKE ALARMS IN SET 1 AND SET 2

The blue portion of the histogram shows the central 95 % of the distribution, while the red portions in the tails of the distribution (and all more extreme values) comprise the least likely 5 % of the distribution (2.5 % in each tail). The fact that a difference of zero falls outside the central 95 % bounds indicates that the average difference in unit response between Model A smoke alarms in Set 1 and Model A smoke alarms in Set 2 is statistically significant. More information on the methodology used to compute this type of confidence interval is given in Appendix A.

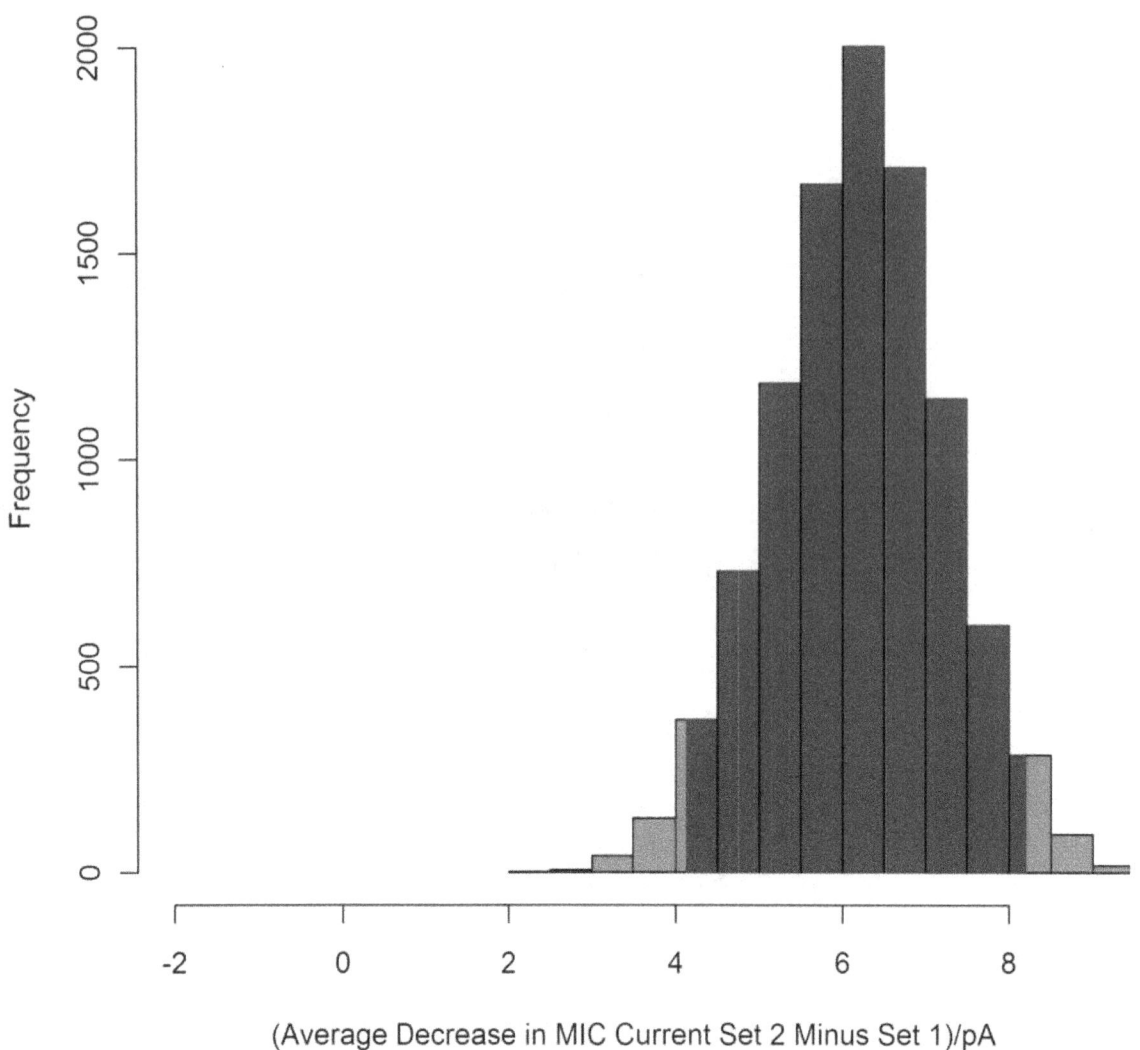

(Average Decrease in MIC Current Set 2 Minus Set 1)/pA

95 % Confidence Interval for Average MIC Current Decrease: (4.1 , 8.2) pA

FIGURE D. 6: HISTOGRAM SHOWING THE DISTRIBUTION OF RESAMPLED ESTIMATES OF THE AVERAGE
DIFFERENCE IN SMOKE ALARM RESPONSE FOR ALL MODEL B SMOKE ALARMS

The blue portion of the histogram shows the central 95 % of the distribution, while the red portions
in the tails of the distribution (and all more extreme values) comprise the least likely 5 % of the
distribution (2.5 % in each tail). The fact that a difference of zero does not fall within the central
95 % bounds indicates that the average difference in unit response between Model B smoke alarms
in Set 1 and Model B smoke alarms in Set 2 is statistically significant. More information on the
methodology used to compute this type of confidence interval is given in Appendix A.

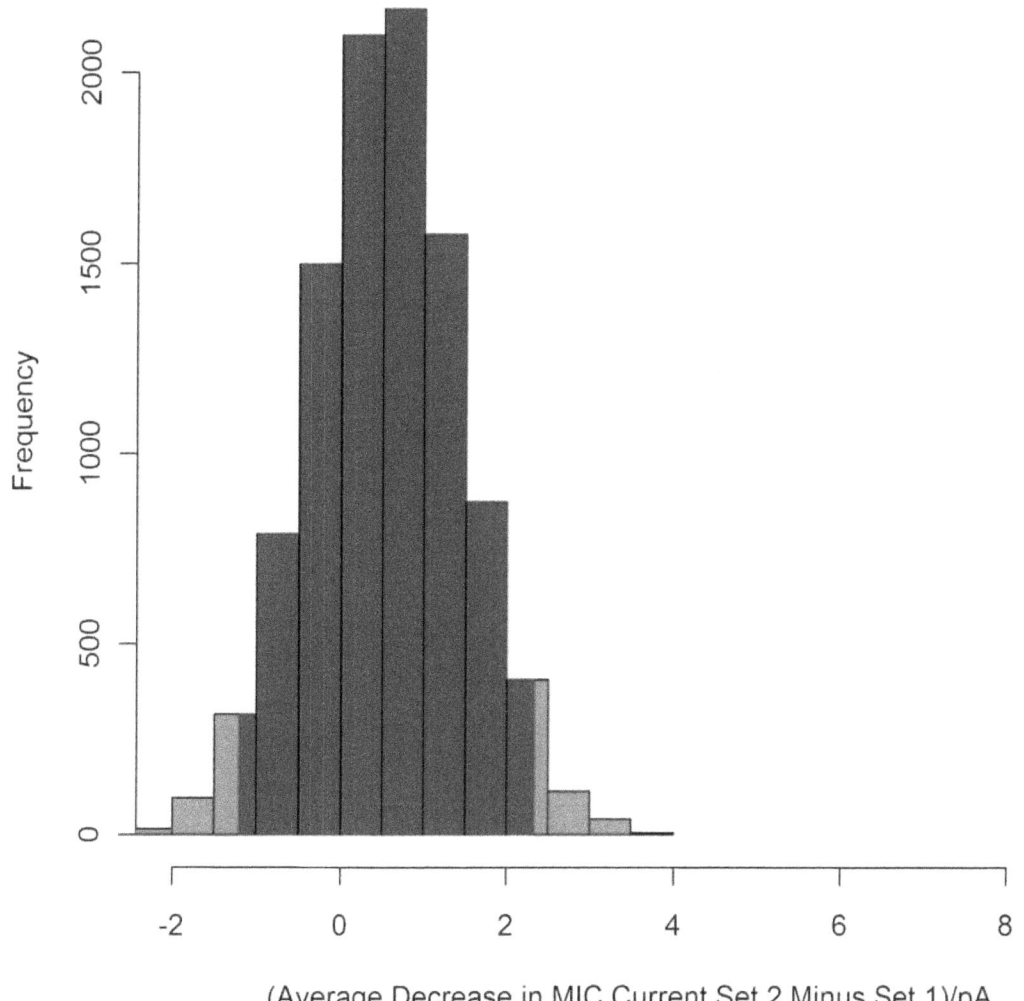

(Average Decrease in MIC Current Set 2 Minus Set 1)/pA

95 % Confidence Interval for Average MIC Current Decrease: (-1.2 , 2.3) pA

FIGURE D. 7: HISTOGRAM SHOWING THE DISTRIBUTION OF RESAMPLED ESTIMATES OF THE AVERAGE DIFFERENCE IN SMOKE ALARM RESPONSE FOR ALL MODELS OF IONIZATION SMOKE ALARMS

This histogram assumes that the distribution of different smoke alarms in use is the same as the distribution of smoke alarm models sampled (approximately 50 % Model A and 25 % each of the other two models). The blue portion of the histogram shows the central 95 % of the distribution, while the red portions in the tails of the distribution (and all more extreme values) comprise the least likely 5 % of the distribution (2.5 % in each tail). The fact that a difference of zero falls inside the central 95 % bounds indicates that the average difference in unit response between the Set 1 and Set 2 smoke alarms is not statistically significant. More information on the methodology used to compute this confidence interval is given in Appendix A.

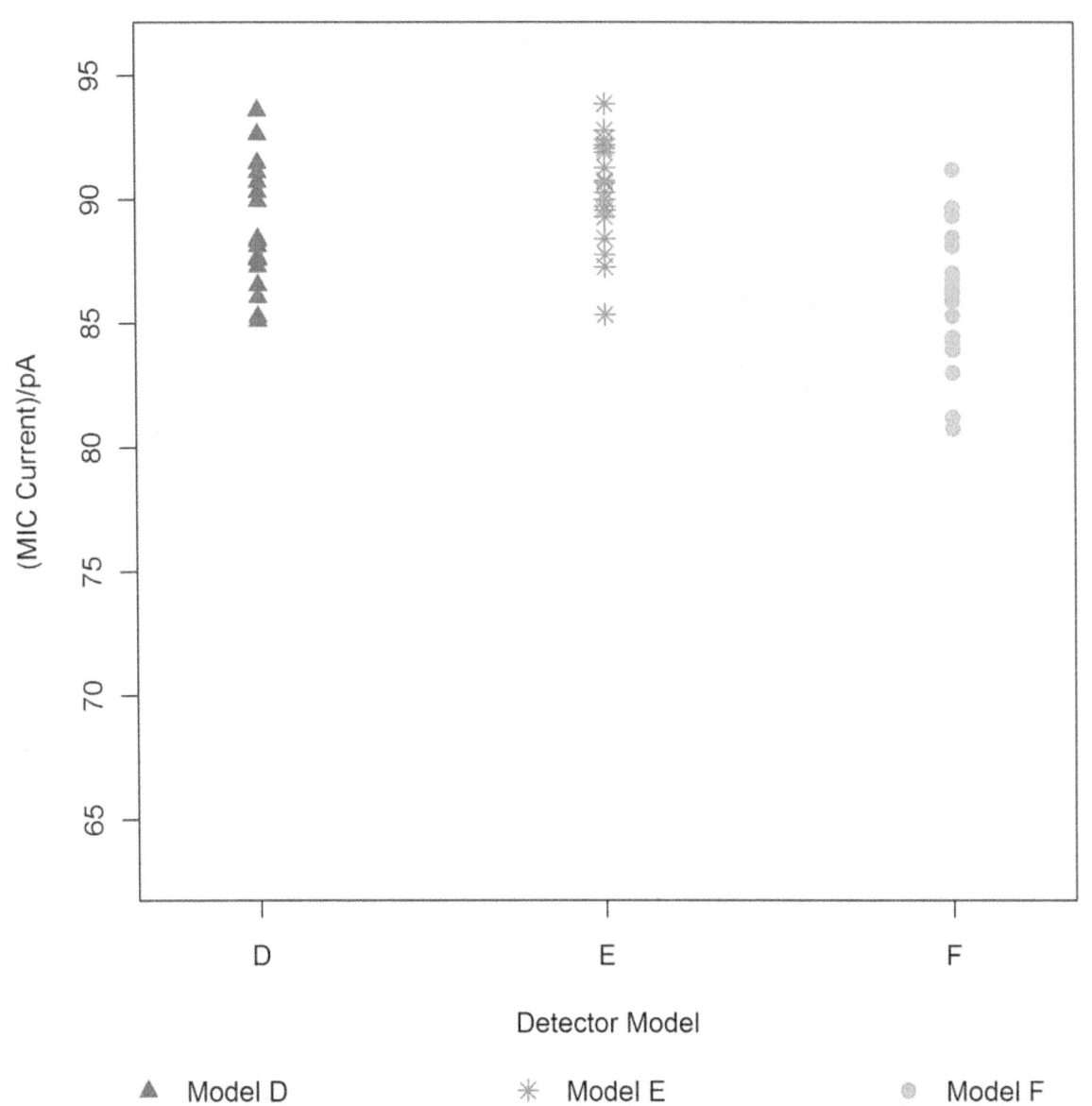

FIGURE E. 1: PLOT OF SMOKE ALARM RESPONSES VS. MODEL FOR SET 3 IONIZATION SMOKE ALARMS.

Responses from multiple tests of six different units from each of three models are shown. The results are color coded by smoke alarm model. Based on these data, one can see that the responses of Models D and E are very similar with respect to both their average responses and their levels of random variation between smoke alarms and measurements. The responses for the Model F smoke alarms appear to be slightly lower on average. The level of random variation in the responses of the Model F smoke alarms looks about the same as for the other two models.

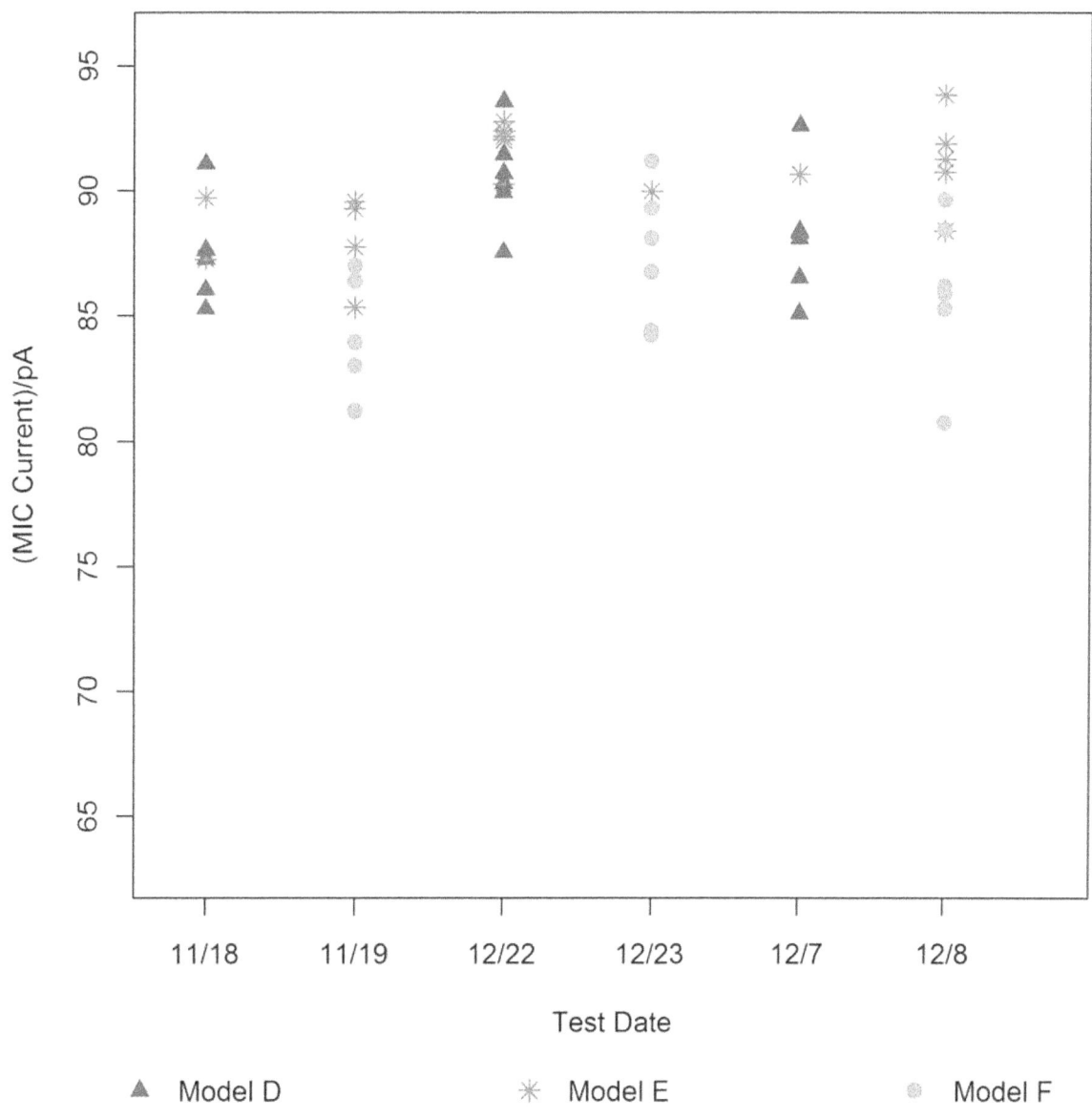

FIGURE E. 2: PLOT OF SET 3 SMOKE ALARM RESPONSES VS. TEST DATE.

The color coding indicates the smoke alarm model. The tests on most dates look similar with respect to both mean response and random variation. Although the results on 11/19 and 12/22 seem to indicate days with low and high results, respectively, this apparent difference is likely a consequence of two factors: no Model F units were tested on 11/19, and the results for Model F units appear to be slightly lower than for the other two models. Within each model, the responses look fairly consistent across all days.

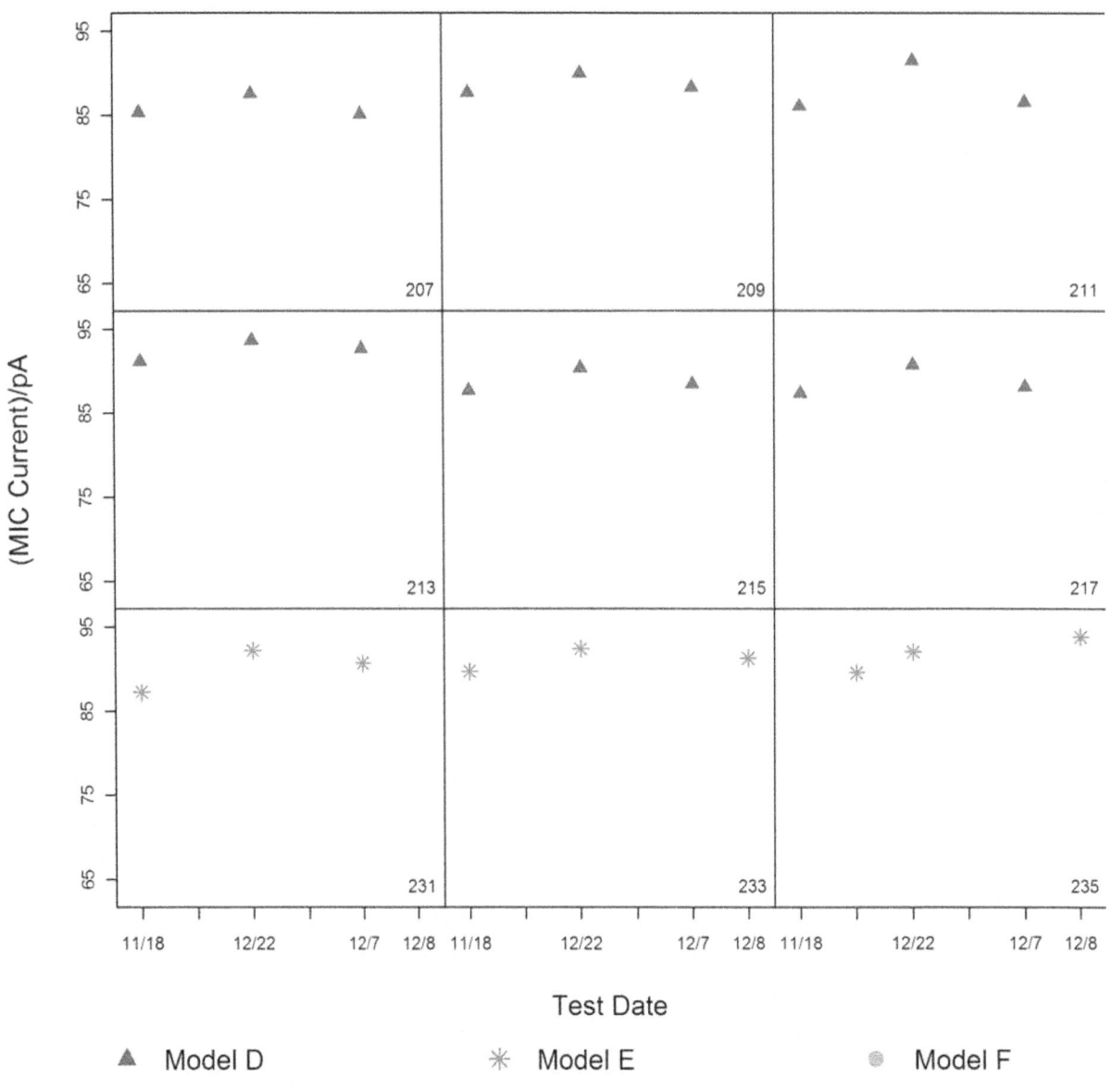

FIGURE E. 3: SET 3 SMOKE ALARM RESPONSE VS. TEST DATE BY UNIT

The smoke alarm models are indicated by the symbol color. The unit identification numbers are given in the lower right corner of each plot. These plots show in more detail how each unit responded to runs made on different dates. Looking across the plots for each model, the responses on different dates look relatively consistent. Comparing the plots for different models, the fact that the responses for Model F (on the following page) are lower is again evident.

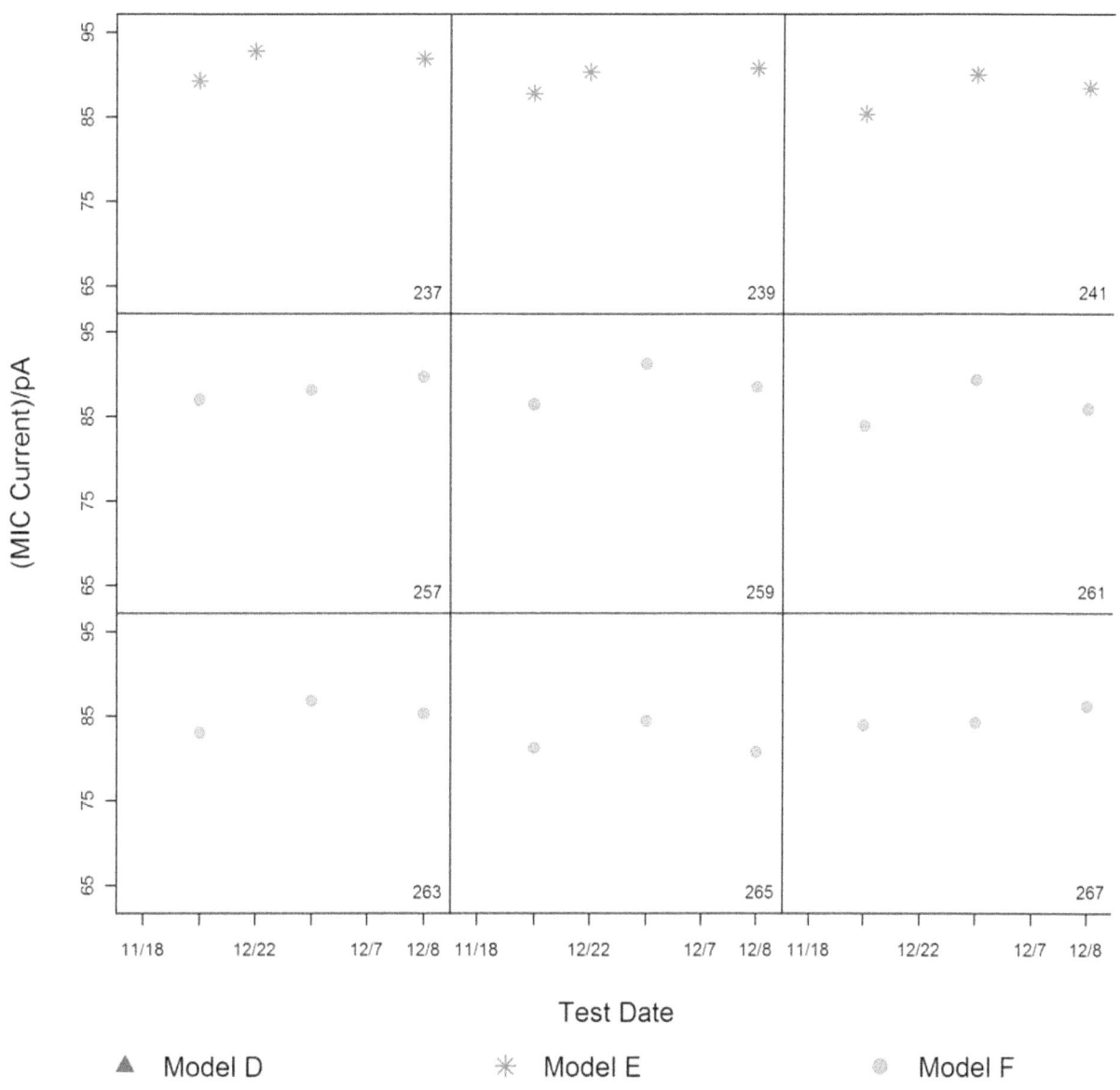

FIGURE E.3 CONTINUED: CAPTION AND INTERPRETATION ON PREVIOUS PAGE.

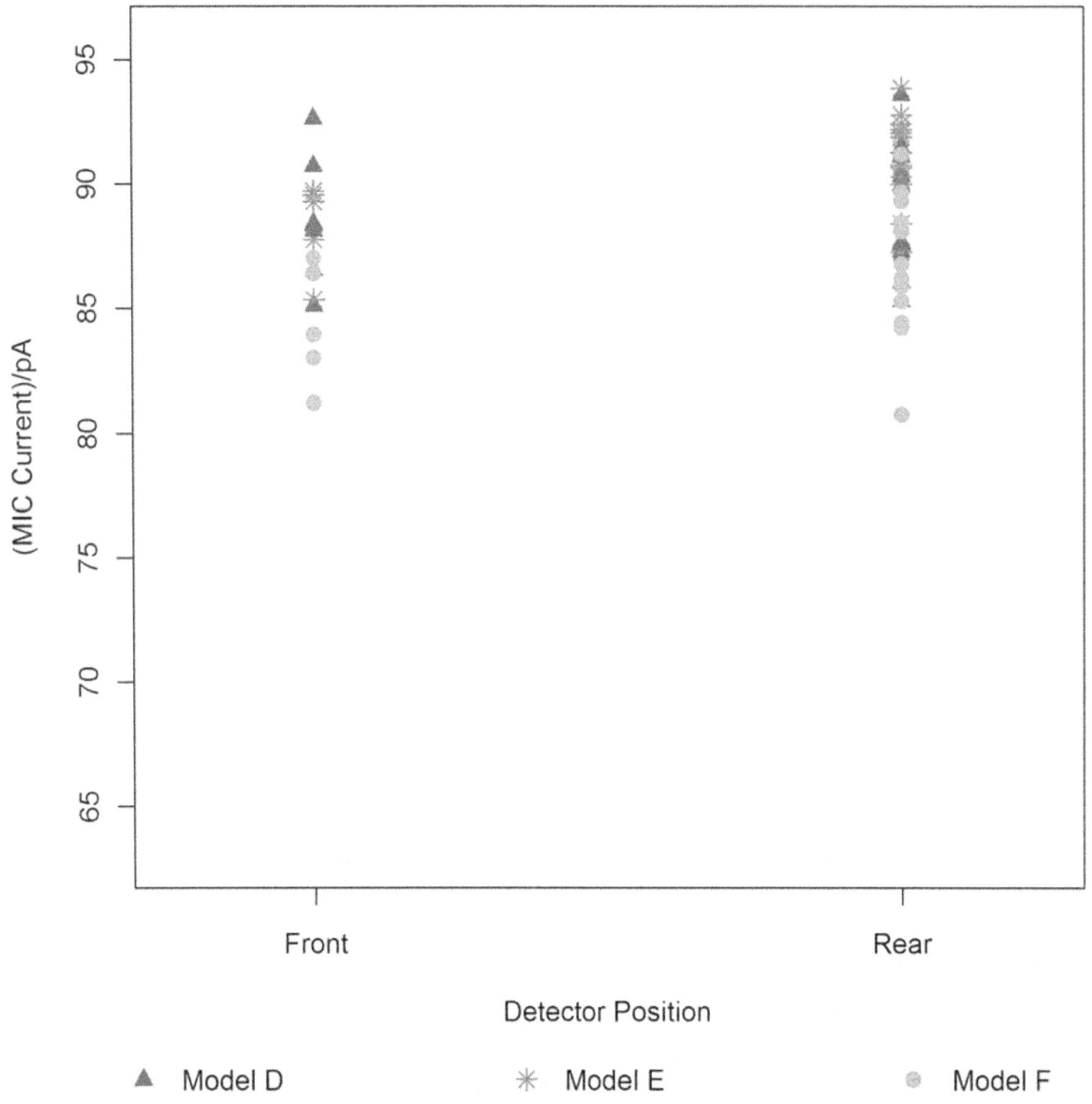

FIGURE E. 4: SET 3 SMOKE ALARM RESPONSES VS. POSITION IN THE FEDE BY MODEL

The smoke alarm model is indicated by the color coding. The similar responses for each position indicate that this factor does not affect the results on average for any of the three smoke alarm models.

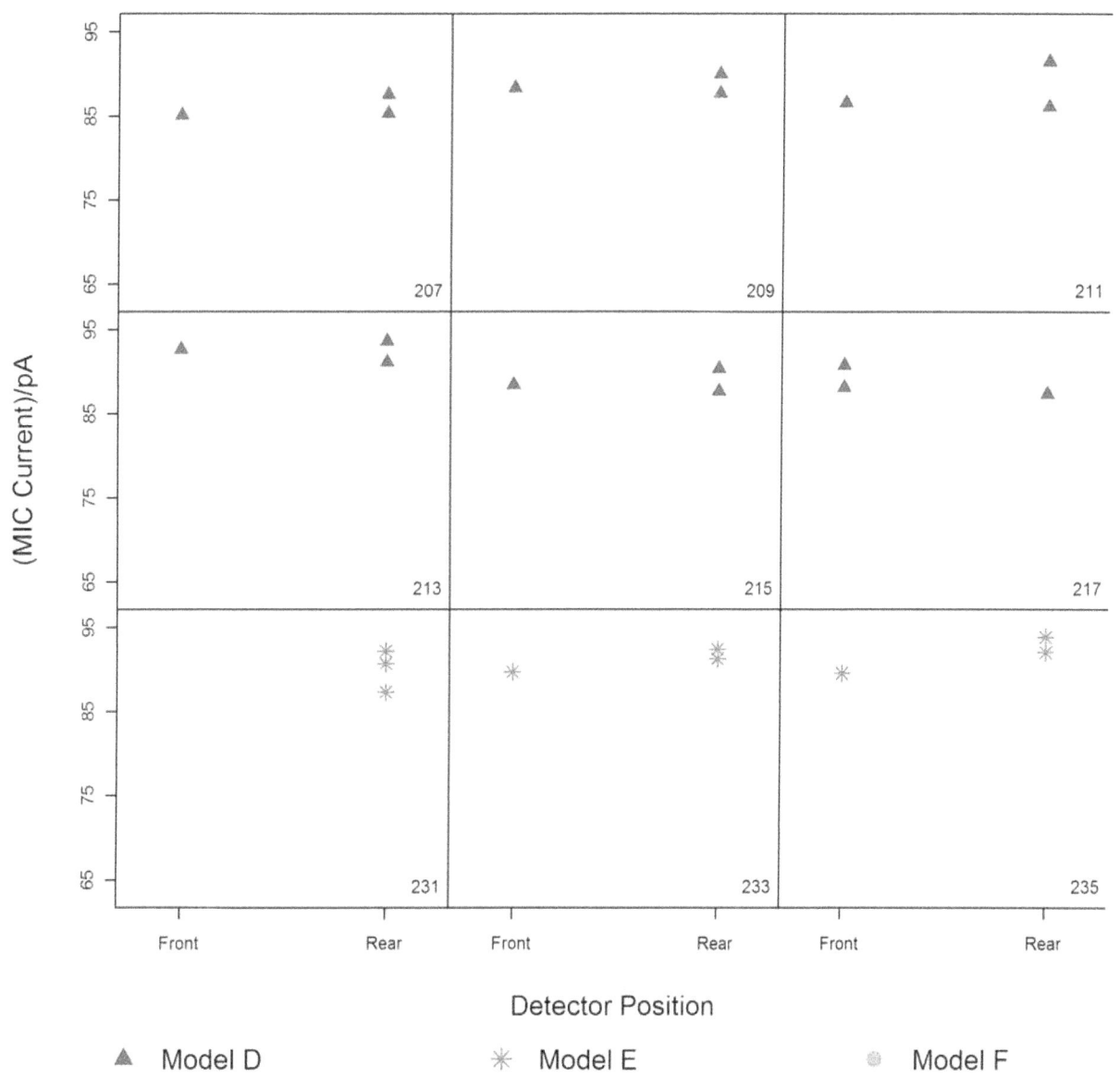

FIGURE E. 5: SET 3 SMOKE ALARM RESPONSES VS. POSITION BY UNIT

The smoke alarm models are indicated by the color coding, and the unit identification numbers are shown in the lower right corner of each plot. These plots provide more detail about individual unit responses to position in the FEDE test apparatus. In all cases, the responses for each unit look relatively consistent, regardless of position.

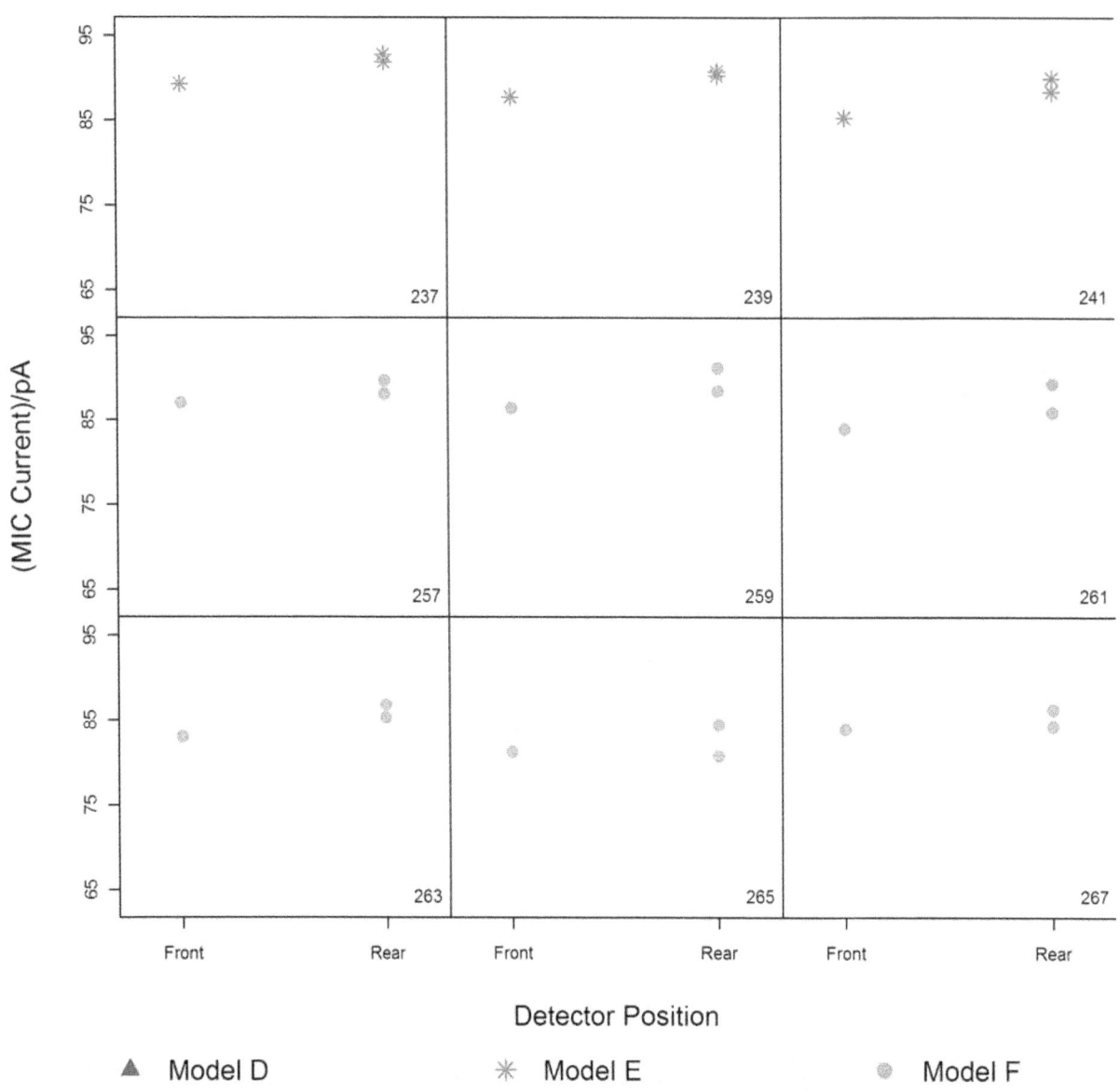

FIGURE E.5 CONTINUED: CAPTION AND INTERPRETATION ON PREVIOUS PAGE

APPENDIX F: EXPLORATORY DATA ANALYSIS FOR SET 4
IONIZATION SMOKE ALARMS

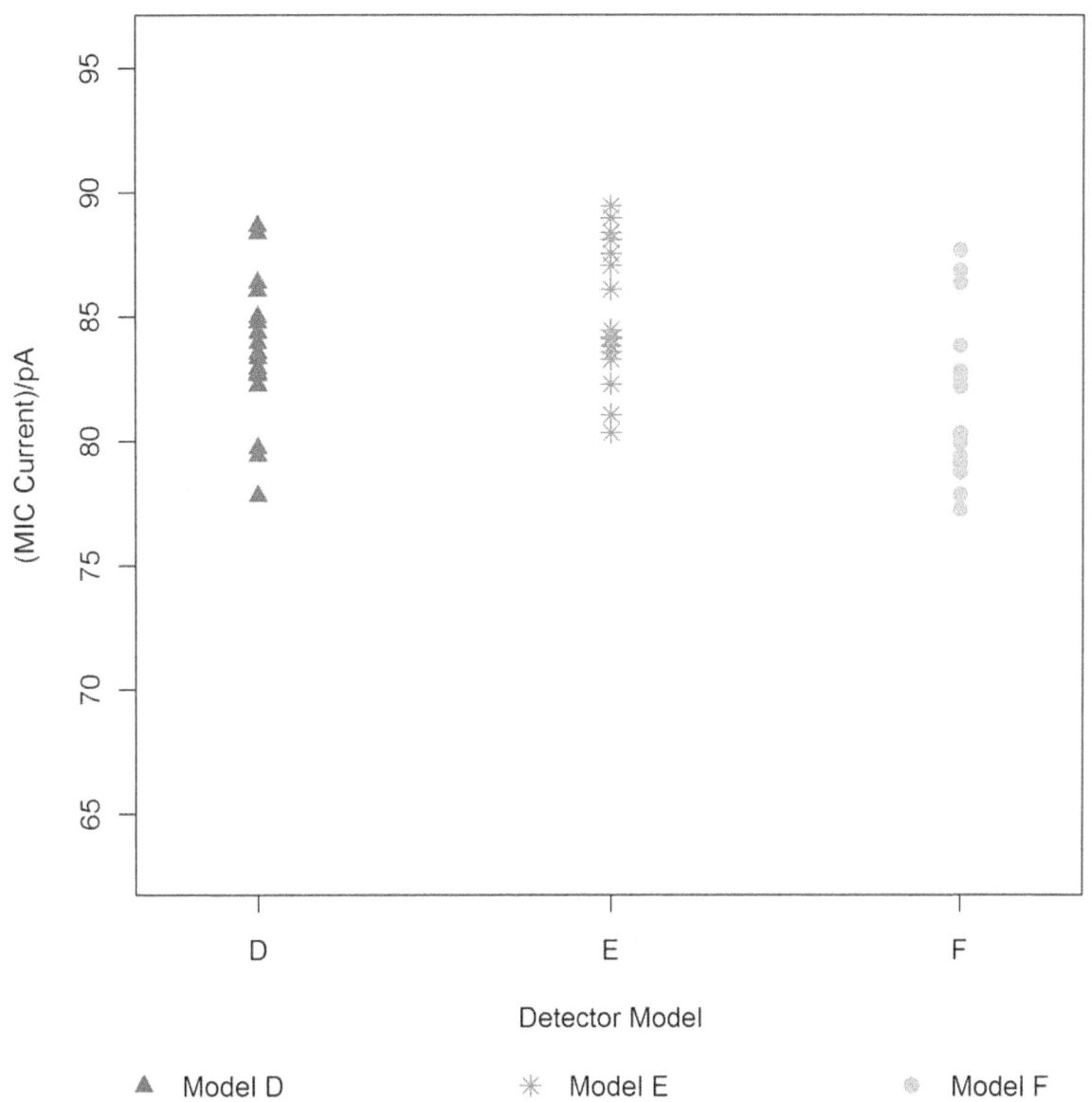

FIGURE F. 1: SMOKE ALARM RESPONSES VS. MODEL FOR SET 4 IONIZATION SMOKE ALARMS

The responses from multiple runs on six different smoke alarms from each of three models are color coded by smoke alarm model. The model-to-model variation appears to be slightly less than that seen when the smoke alarms were new, although the MIC values are not strikingly different and may not reflect any real change caused by the accelerated aging. The scatter in the data looks essentially like the scatter observed when the units were new as well.

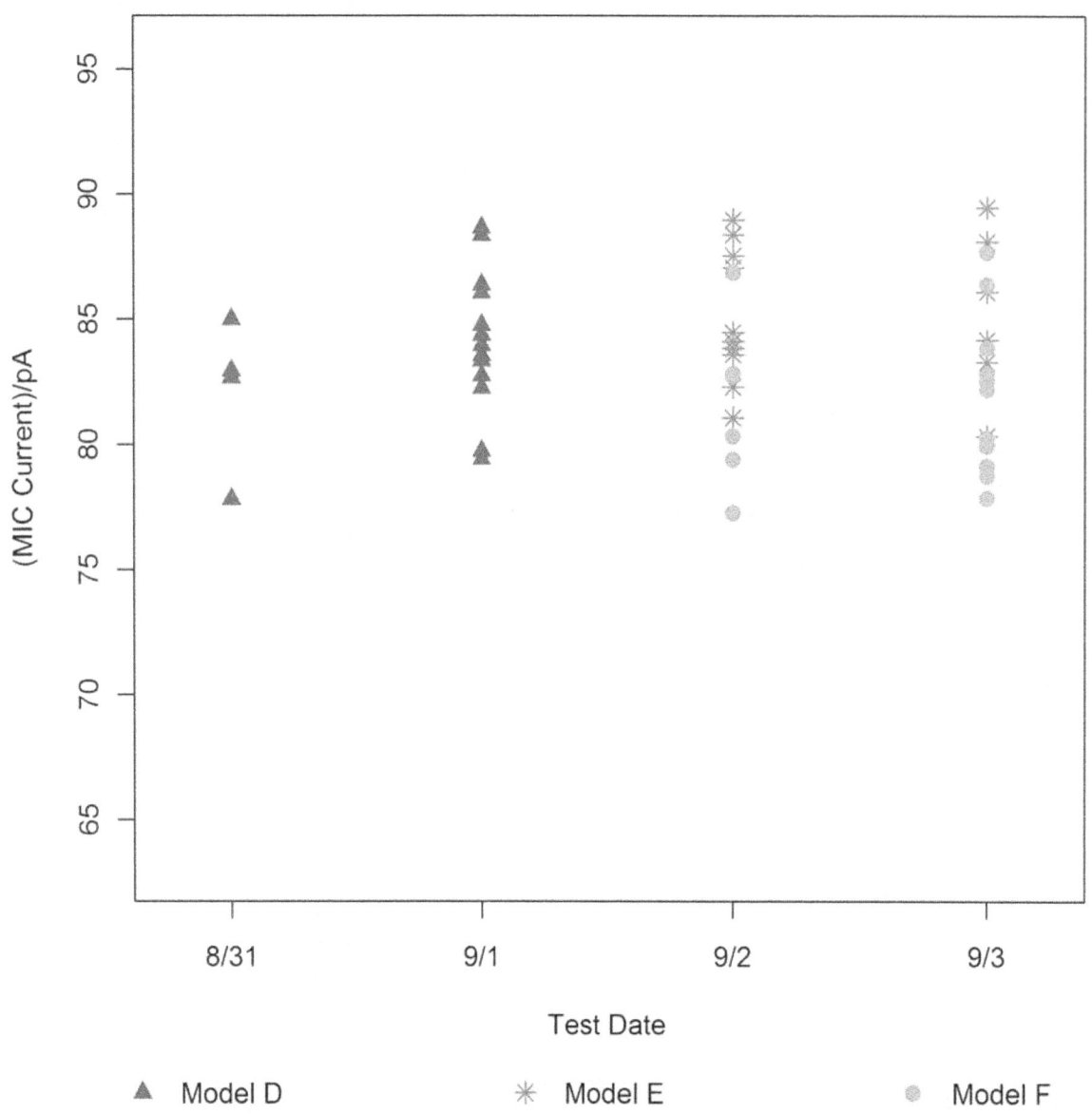

FIGURE F. 2: SMOKE ALARM RESPONSES VS. TEST DATE FOR SET 4 IONIZATION SMOKE ALARMS

The color coding indicates the smoke alarm model. The measurements made on different days look consistent across days for all three models. Seeing the data for Models E and F plotted on top of one another suggests that the results for Model F may be slightly lower than for Model E, as seen when the units were new, and in contrast to Figure F.1.

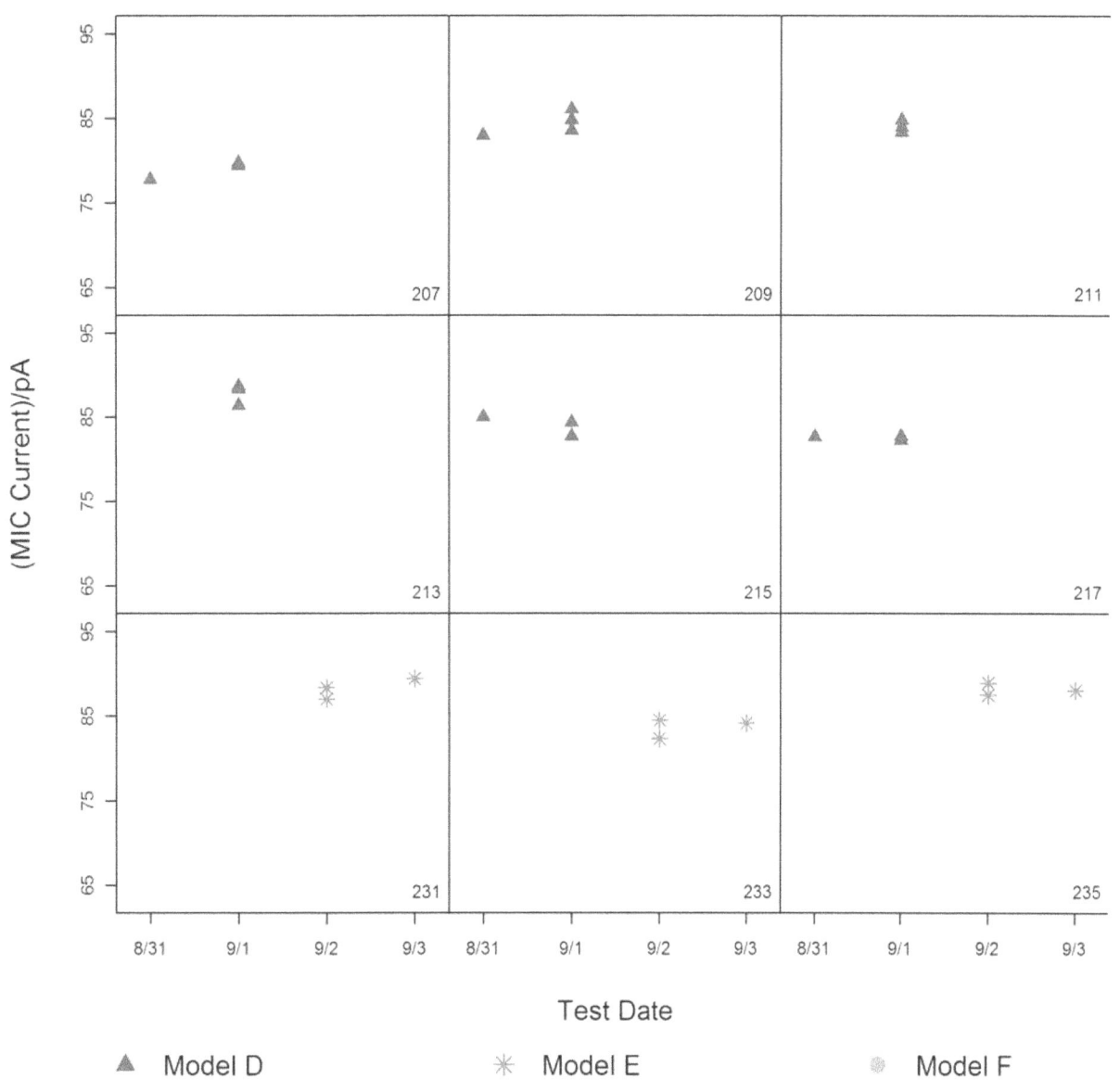

FIGURE F. 3: SET 4 SMOKE ALARM RESPONSE VS. TEST DATE BY UNIT

The smoke alarm models are indicated by the symbol color. The unit identification numbers are given in the lower right corner of each plot. For these tests, each unit was tested on only one day (and in one position in the FEDE). However, looking across the plots for each model, the responses for different units look quite consistent overall.

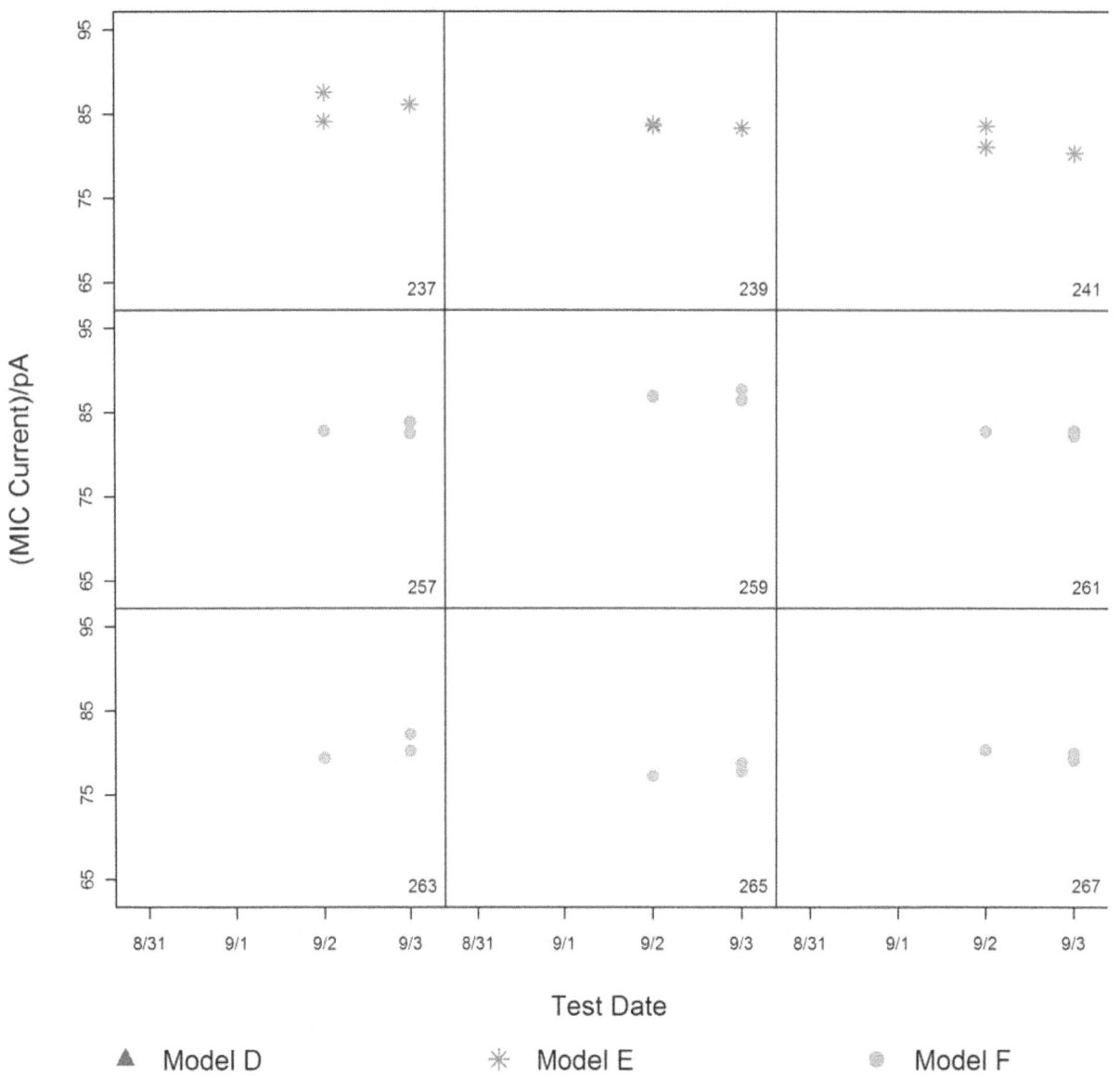

FIGURE F.3 CONTINUED: CAPTION AND INTERPRETATION ON PREVIOUS PAGE.

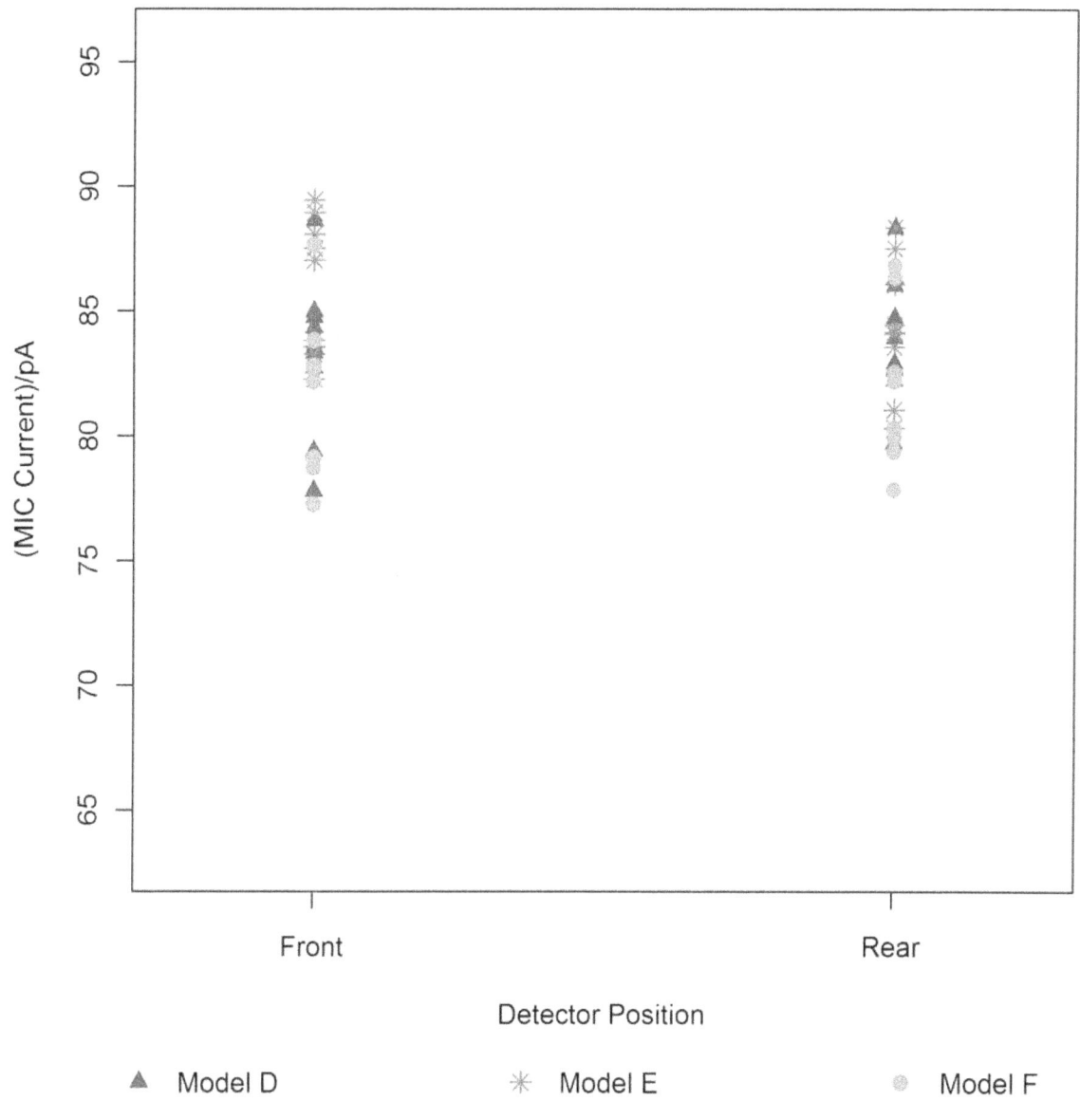

FIGURE F. 4: SET 4 SMOKE ALARM RESPONSES VS. POSITION IN THE FEDE

The smoke alarm model is indicated by the color coding. The similar responses for each position indicate that this factor does not affect the results on average for any of the three models.

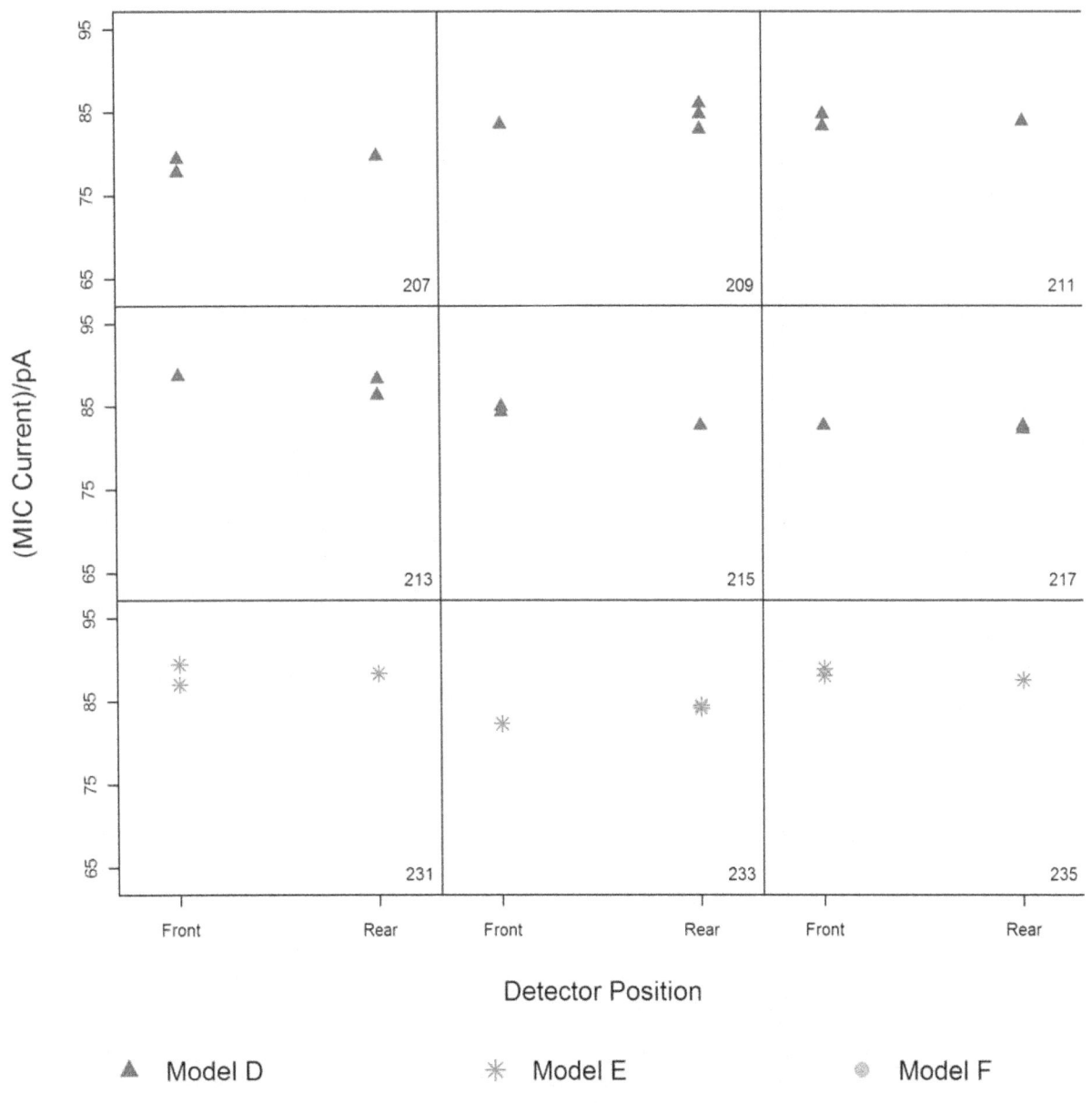

FIGURE F. 5: SET 4 SMOKE ALARM RESPONSES VS. POSITION BY UNIT

The smoke alarm models are indicated by the color coding, and the unit numbers are shown in the lower right corner of each plot. These plots provide more detail about individual unit responses. In all cases, the responses across smoke alarms look relatively consistent for each model, regardless of position.

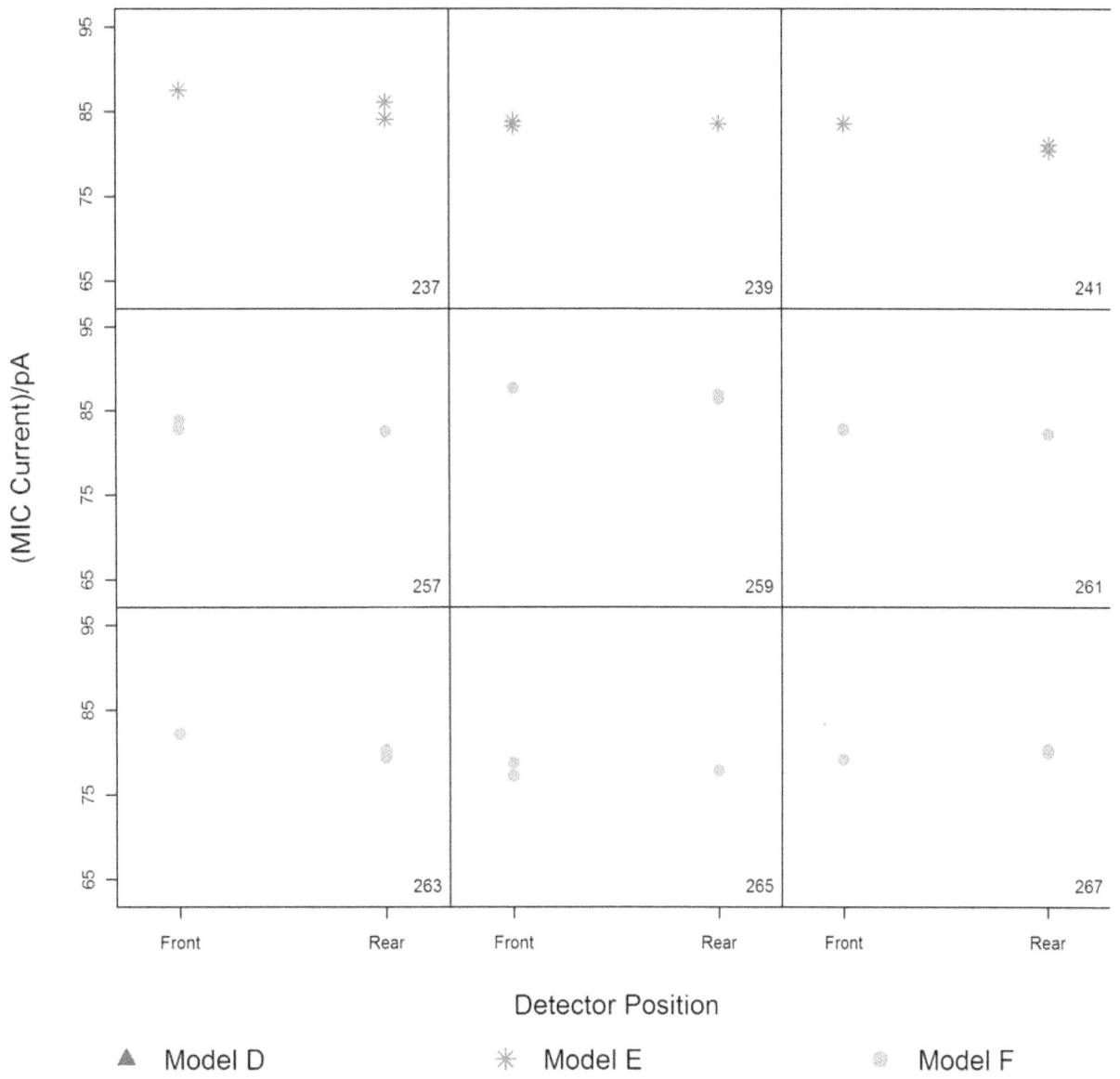

FIGURE F.5 CONTINUED: CAPTION AND INTERPRETATION ON PREVIOUS PAGE

67

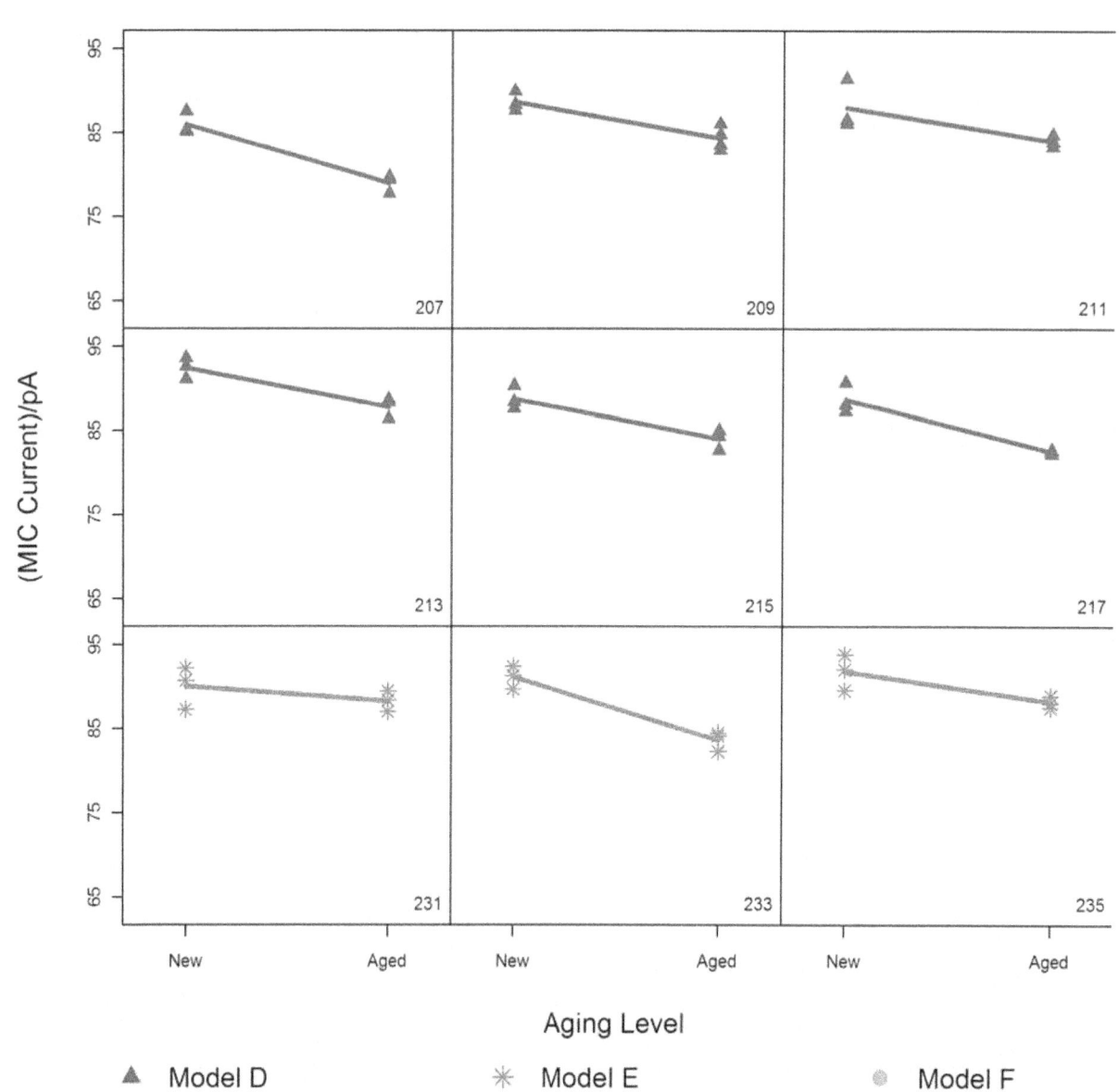

FIGURE G. 1: SET 3 AND SET 4 SMOKE ALARM RESPONSES VS. AN INDICATOR OF SMOKE ALARM AGING BY
UNIT

The responses from multiple tests of six different units from each of three models are color coded
by smoke alarm model. The unit identification numbers are shown in the lower right corner of each
plot. The diagonal lines on each plot, which connect the mean responses observed before and after
aging, are included to facilitate determining if the observed response for each unit increased or
decreased after aging. These plots show that, for every unit tested, the MIC was lower after aging
than it had been when the unit was new. There was a general decrease in smoke alarm sensitivity.

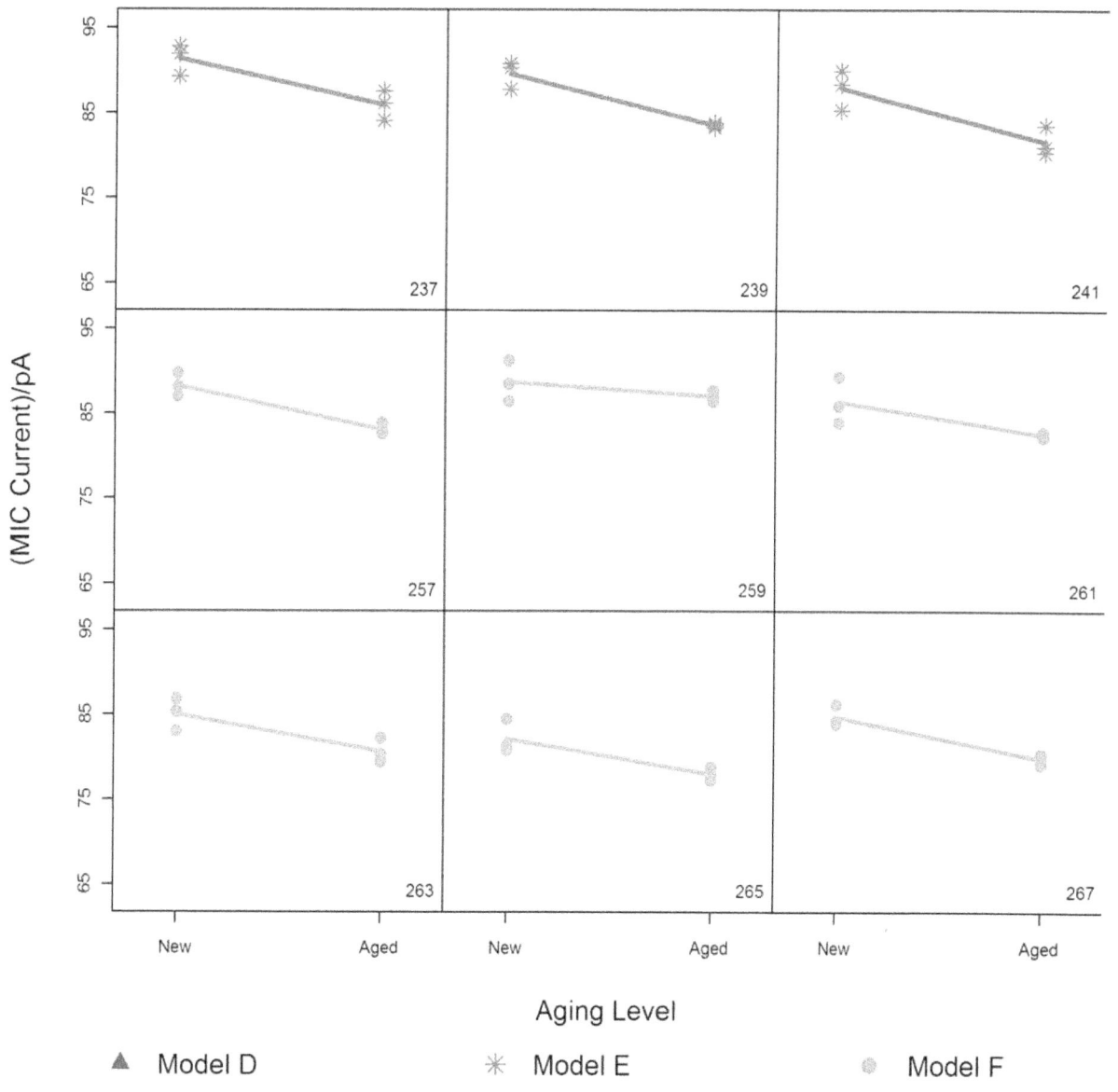

FIGURE G.1 CONTINUED: CAPTION AND INTERPRETATION ON PREVIOUS PAGE.

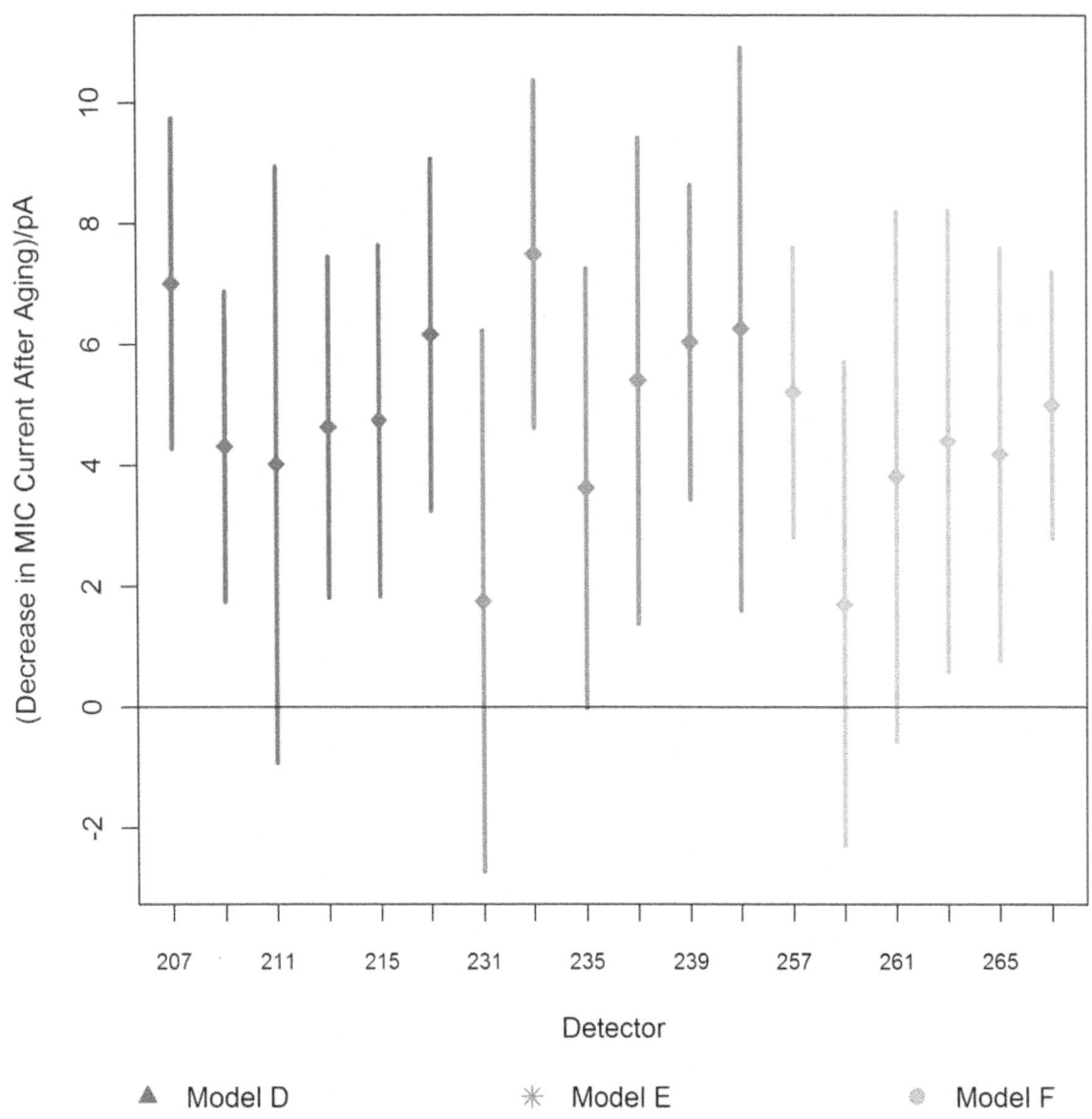

FIGURE G. 2: INDIVIDUAL 95 % CONFIDENCE INTERVALS (CI) FOR THE AVERAGE DIFFERENCE IN RESPONSE FOR EACH SMOKE ALARM BETWEEN SET 3 AND SET 4

The results are color-coded by smoke alarm model. Significant differences in response before and after aging are indicated by intervals that do not contain zero (*e.g.*, smoke alarms 207 and 209). The fact that a difference in response for any individual unit is not significant does not mean there is no difference, just that any difference that exists cannot be detected with the amount of data for that unit. This supports the finding expressed in the interpretation of Figure G.1.

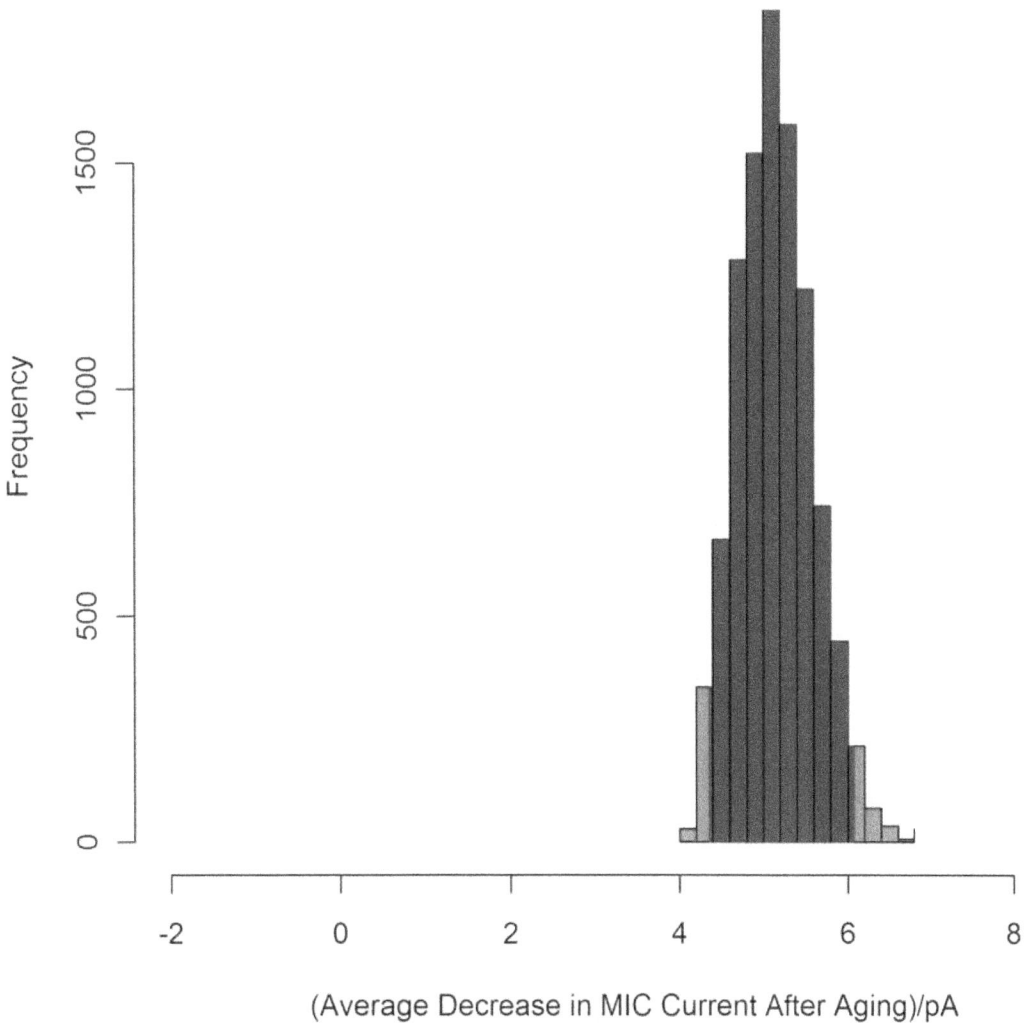

(Average Decrease in MIC Current After Aging)/pA

95 % Confidence Interval for Average MIC Current Decrease: (4.4 , 6.1) pA

FIGURE G. 3: HISTOGRAM SHOWING THE DISTRIBUTION OF RESAMPLED ESTIMATES OF THE AVERAGE
DIFFERENCE IN UNIT RESPONSE FOR ALL SET 3 AND SET 4 MODEL D IONIZATION SMOKE ALARMS

The blue portion of the histogram shows the central 95 % of the distribution, while the red portions
in the tails of the distribution (and all more extreme values) comprise the least likely 5 % of the
distribution (2.5 % in each tail). The fact that a difference of zero falls (well) outside the central
95 % bounds indicates that the average difference in unit response before and after aging is
statistically significant. More information on the methodology used to compute this confidence
interval is given in Appendix A.

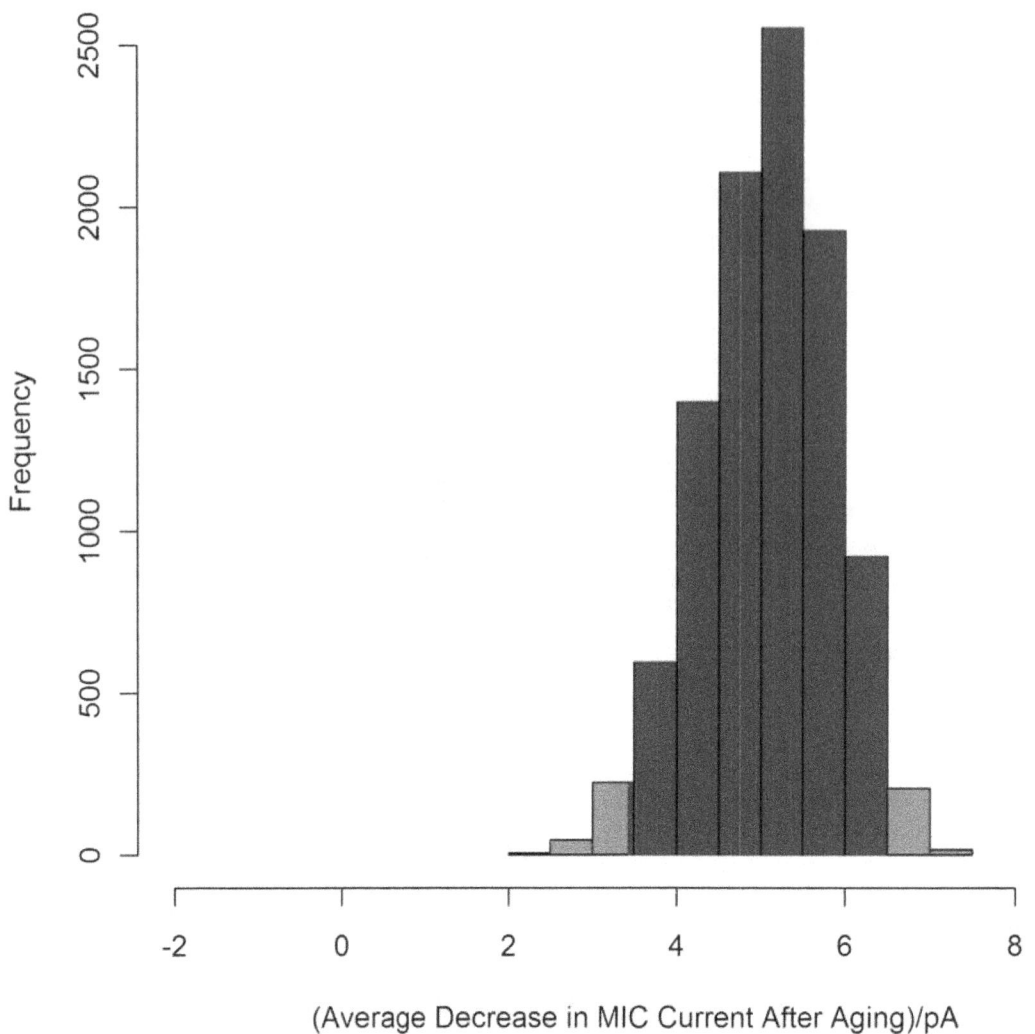

(Average Decrease in MIC Current After Aging)/pA

95 % Confidence Interval for Average MIC Current Decrease: (3.4 , 6.5) pA

FIGURE G. 4: HISTOGRAM SHOWING THE DISTRIBUTION OF RESAMPLED ESTIMATES OF THE AVERAGE
DIFFERENCE IN UNIT RESPONSE FOR ALL SET 3 AND SET 4 MODEL E IONIZATION SMOKE ALARMS

The blue portion of the histogram shows the central 95 % of the distribution, while the red portions
in the tails of the distribution (and all more extreme values) comprise the least likely 5 % of the
distribution (2.5 % in each tail). The fact that a difference of zero falls outside the central 95 %
bounds indicates that the average difference in unit response before and after aging is statistically
significant. More information on the methodology used to compute this type of confidence interval
is given in Appendix A.

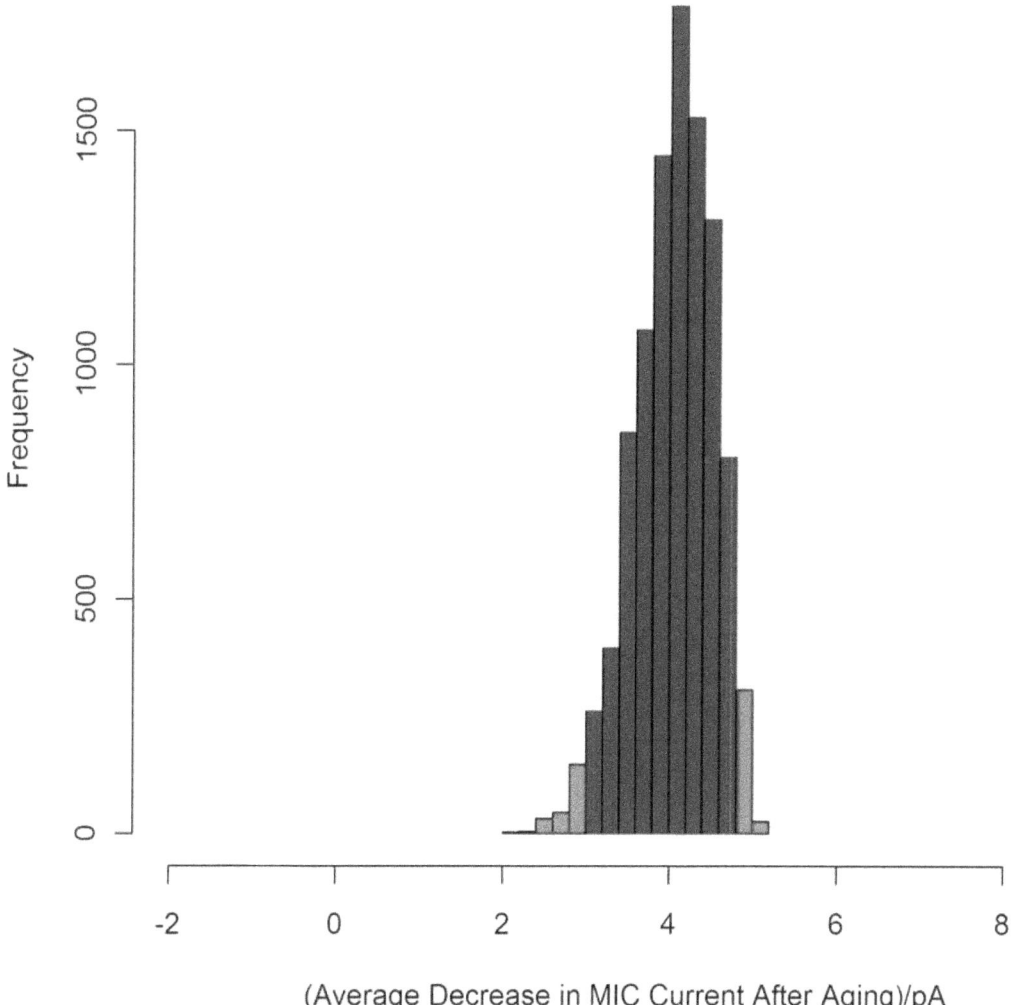

(Average Decrease in MIC Current After Aging)/pA

95 % Confidence Interval for Average MIC Current Decrease: (3 , 4.8) pA

FIGURE G. 5: HISTOGRAM SHOWING THE DISTRIBUTION OF RESAMPLED ESTIMATES OF THE AVERAGE
DIFFERENCE IN UNIT RESPONSE FOR ALL SET 3 AND SET 4 MODEL F IONIZATION SMOKE ALARMS

The blue portion of the histogram shows the central 95 % of the distribution, while the red portions
in the tails of the distribution (and all more extreme values) comprise the least likely 5 % of the
distribution (2.5 % in each tail). In this case, zero does not fall within the central 95 % bounds,
which indicates that the average difference in unit response before and after aging is statistically
significant. More information on the methodology used to compute this type of confidence interval
is given in Appendix A.

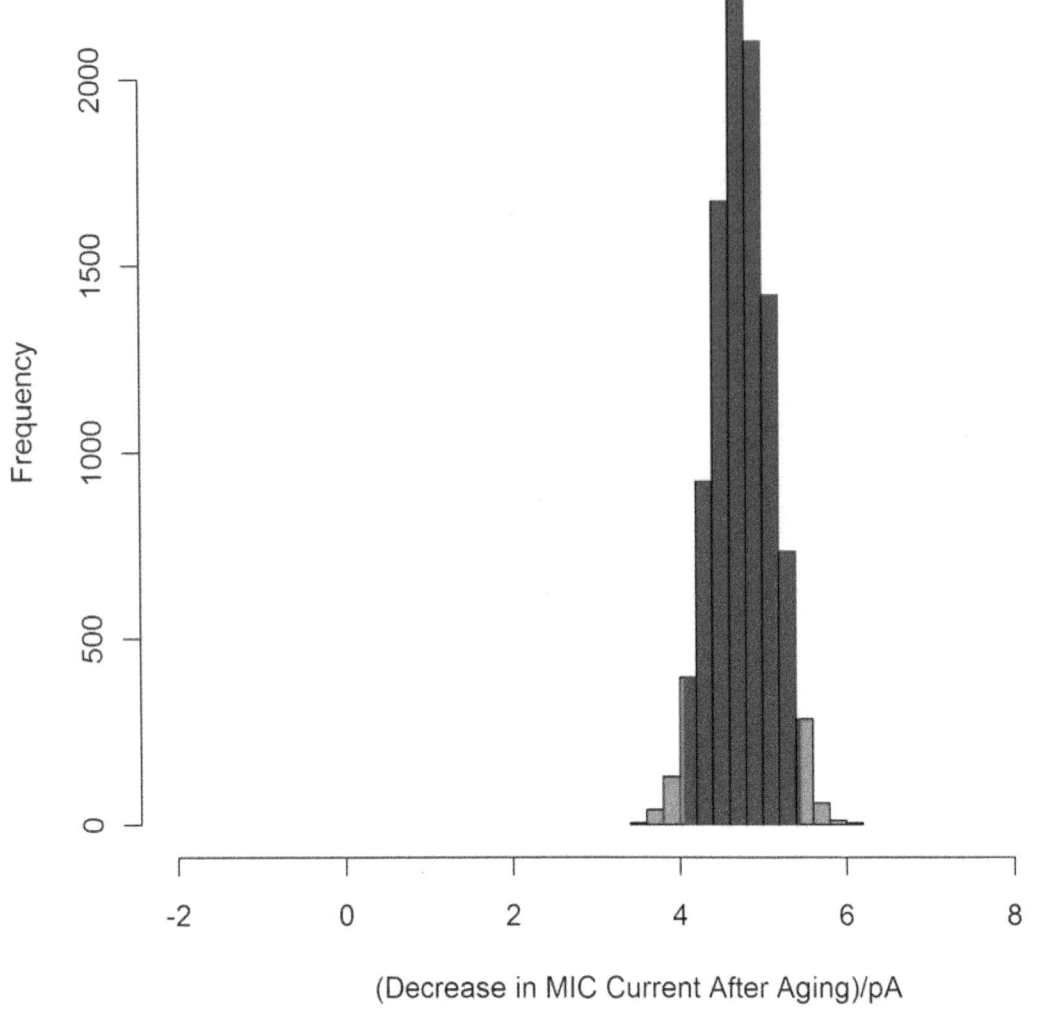

(Decrease in MIC Current After Aging)/pA

95 % Confidence Interval for Average MIC Current Decrease: (4.1 , 5.4) pA

FIGURE G. 6: HISTOGRAM SHOWING THE DISTRIBUTION OF RESAMPLED ESTIMATES OF THE AVERAGE DIFFERENCE IN UNIT RESPONSE FOR ALL MODELS OF SET 3 AND SET 4 IONIZATION SMOKE ALARMS

This histogram assumes that the distribution of different smoke alarms in use is the same as the distribution of smoke alarm models sampled (approximately 50 % Model A and 25 % each of the other two models). The blue portion of the histogram shows the central 95 % of the distribution, while the red portions in the tails of the distribution (and all more extreme values) comprise the least likely 5 % of the distribution (2.5 % in each tail). The fact that a difference of zero falls outside the central 95 % bounds indicates that the average difference in unit response before and after aging is statistically significant. More information on the methodology used to compute this confidence interval is given in Appendix A.

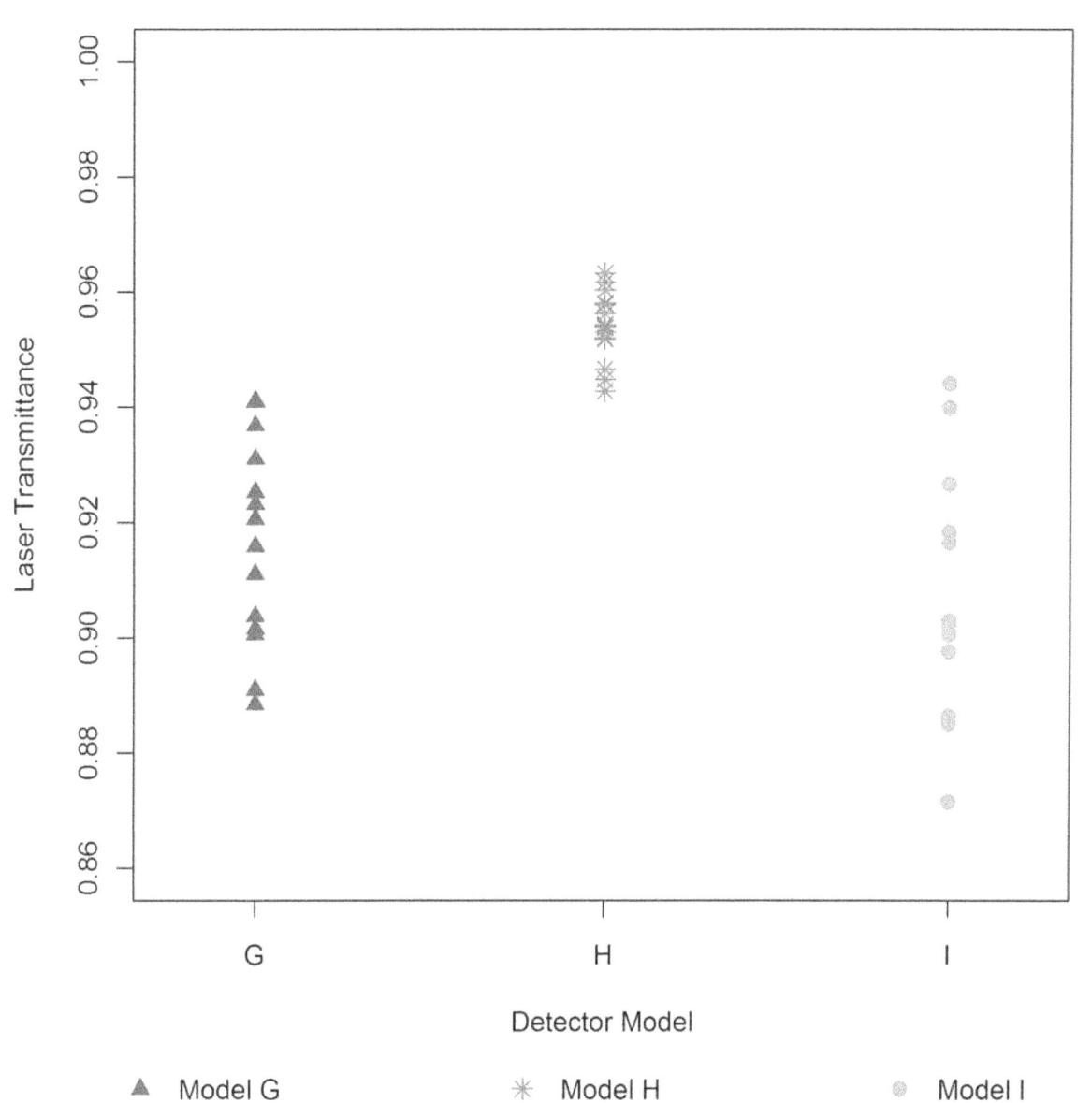

FIGURE H. 1: SMOKE ALARM RESPONSES VS. MODEL FOR SET 3 PHOTOELECTRIC SMOKE ALARMS

The responses from multiple tests of six different units from each of three models are color coded by smoke alarm model. Based on these data, one can see that the responses of Models G and I are similar with respect to both the average response and the random variation between units and measurements. In contrast, the responses from the Model H smoke alarms are higher on average, and less variable.

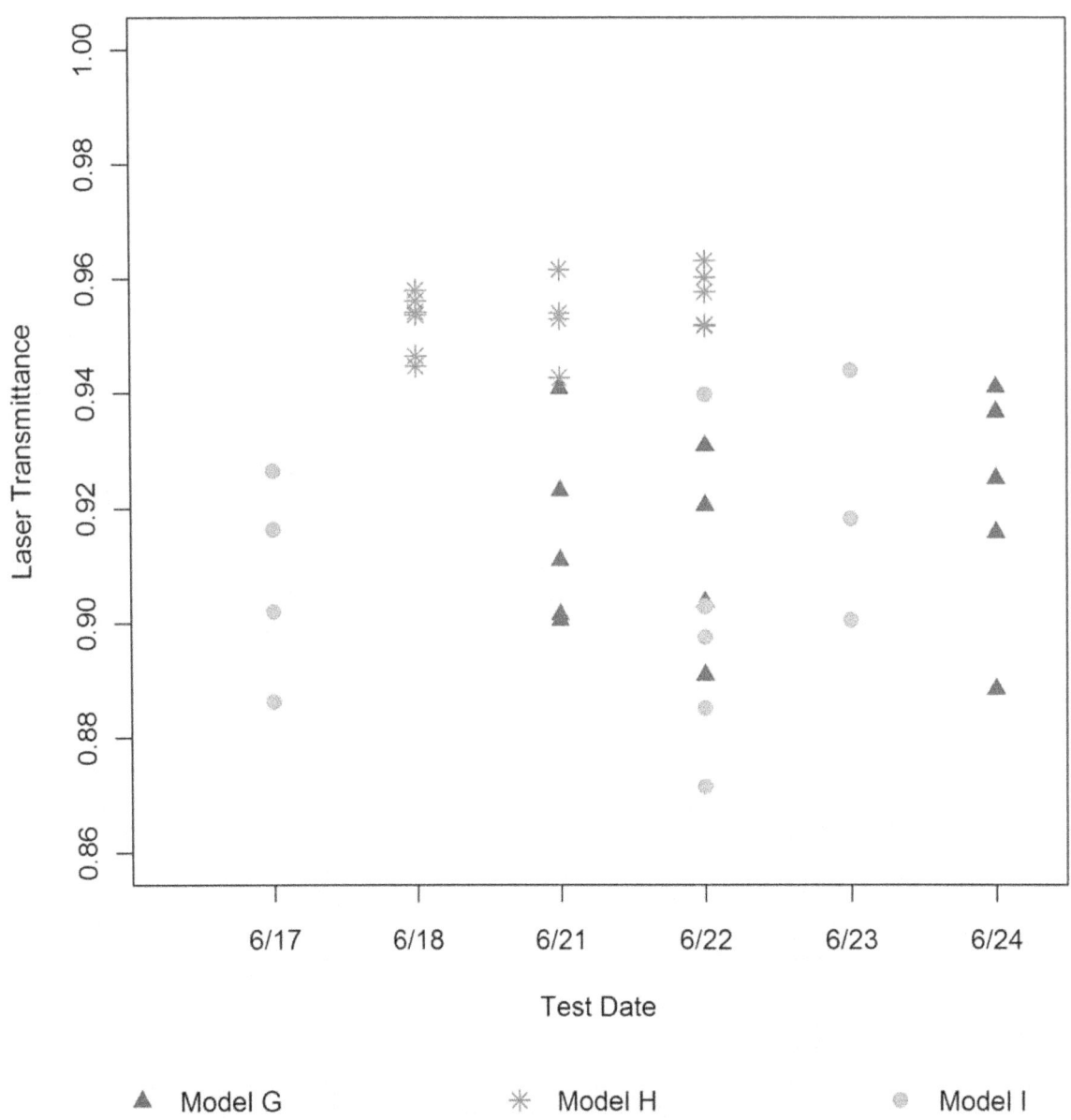

FIGURE H. 2: SET 3 PHOTOELECTRIC SMOKE ALARM RESPONSES VS. TEST DATE

The color coding of the symbols indicates the smoke alarm model. The tests performed on most dates look similar with respect to mean response and random variation. Although the results on June 17 and June 18 seem to indicate days with low and high results, respectively, this apparent difference is simply a consequence of the fact that only units of Models G and H were tested on those days, and the Model H responses are higher than the responses observed for Models G and I, as indicated by the similar responses for Model H smoke alarms on June 21 and June 22. Within each model, the responses look consistent across all days.

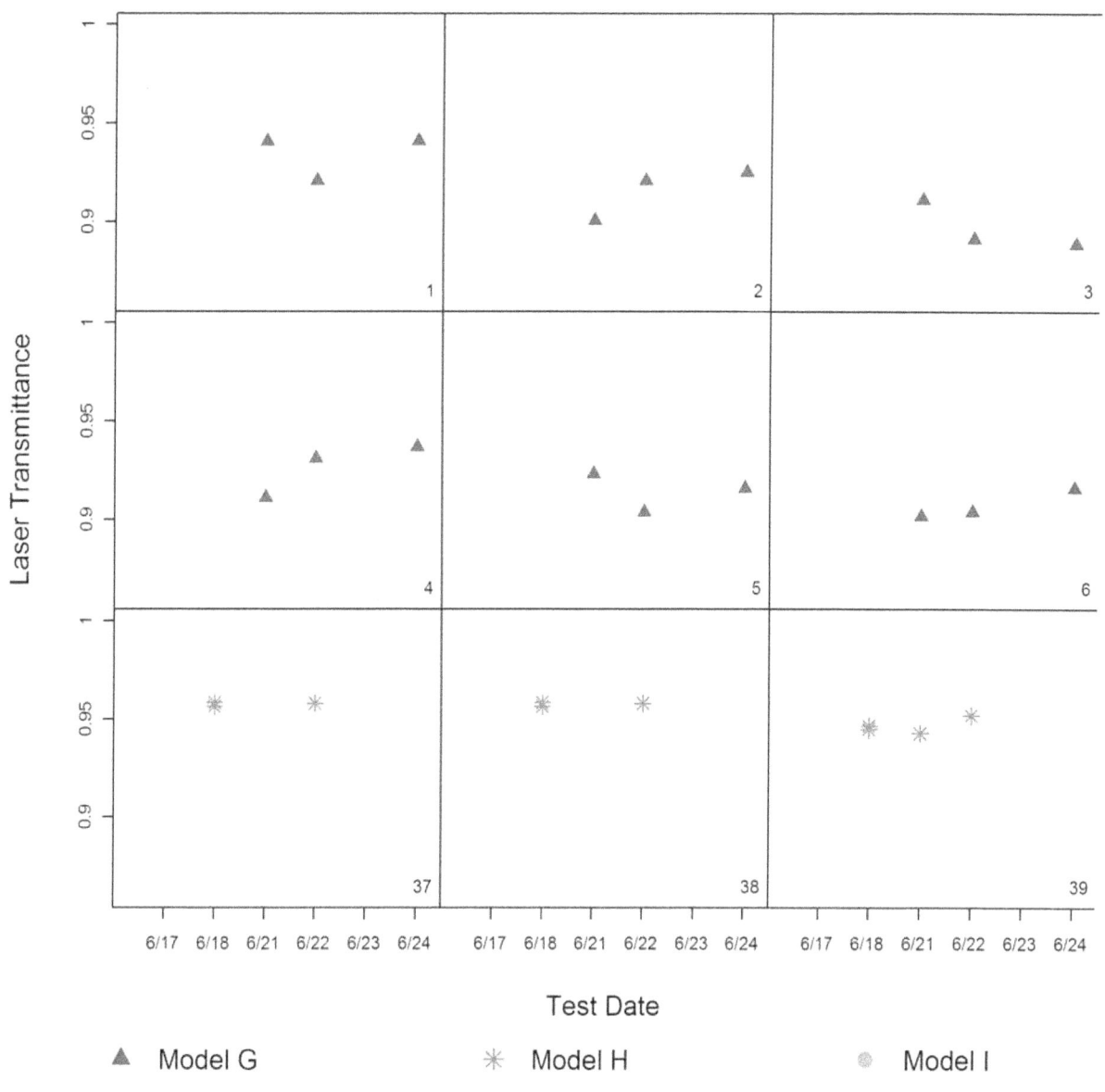

FIGURE H. 3: SMOKE ALARM RESPONSE VS. DATE BY UNIT FOR SET 3 PHOTOELECTRIC SMOKE ALARMS

The smoke alarm models are indicated by the symbol color. The unit identification numbers are given in the lower right corner of each plot. These plots show in more detail how each unit responded to tests performed on different dates. Looking across the plots for each model, the responses on different dates look relatively consistent. Comparing the plots for different models, the fact that the responses for Model H are higher and less dispersed is again evident.

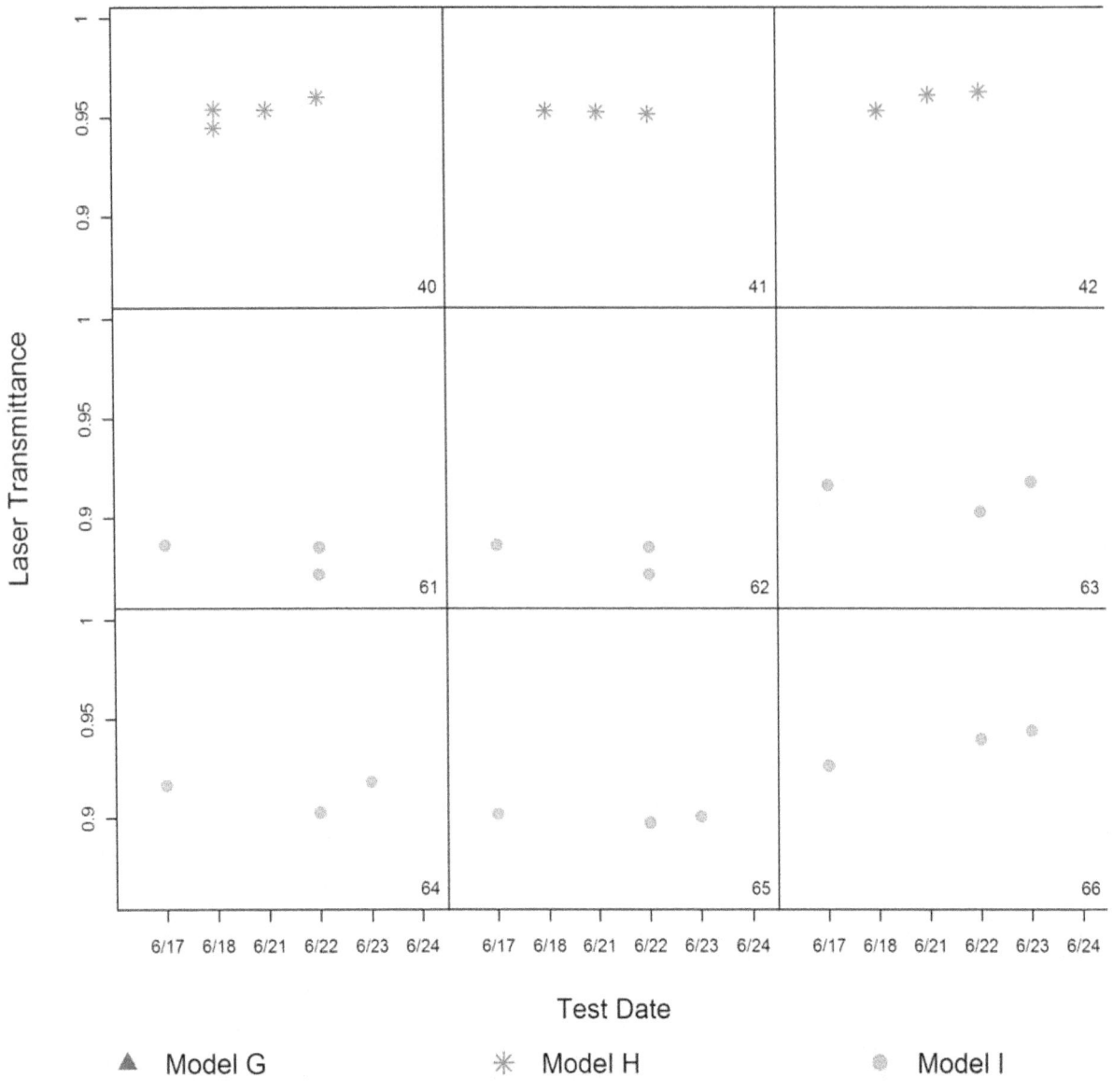

FIGURE H.3 CONTINUED: CAPTION AND INTERPRETATION ON PREVIOUS PAGE

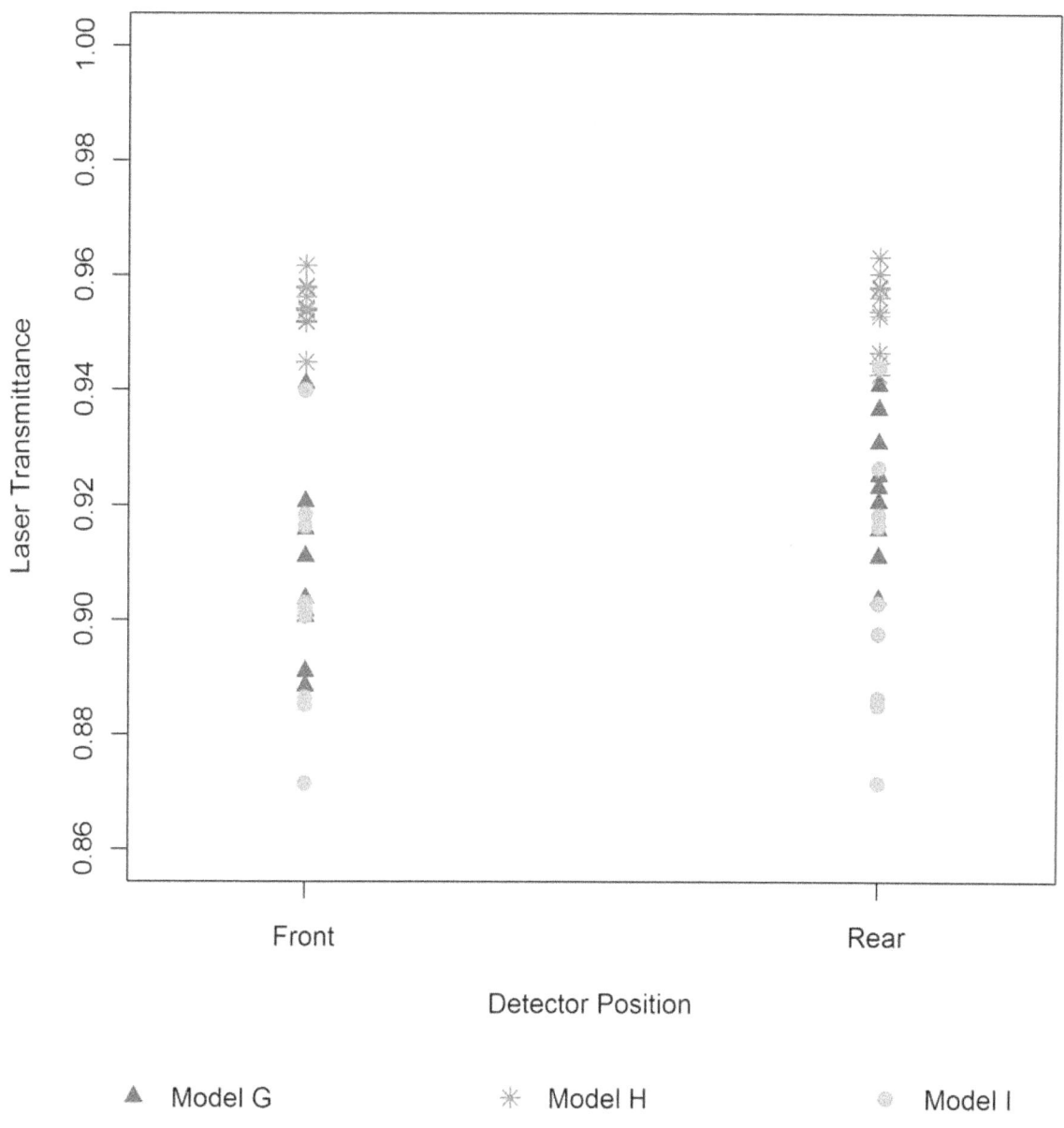

FIGURE H. 4: SET 3 PHTOELECTRIC UNIT RESPONSES VS. POSITION IN THE FEDE

The smoke alarm model is indicated by the color coding. The similar responses for each position indicate that this factor does not affect the results on average for any of the three models.

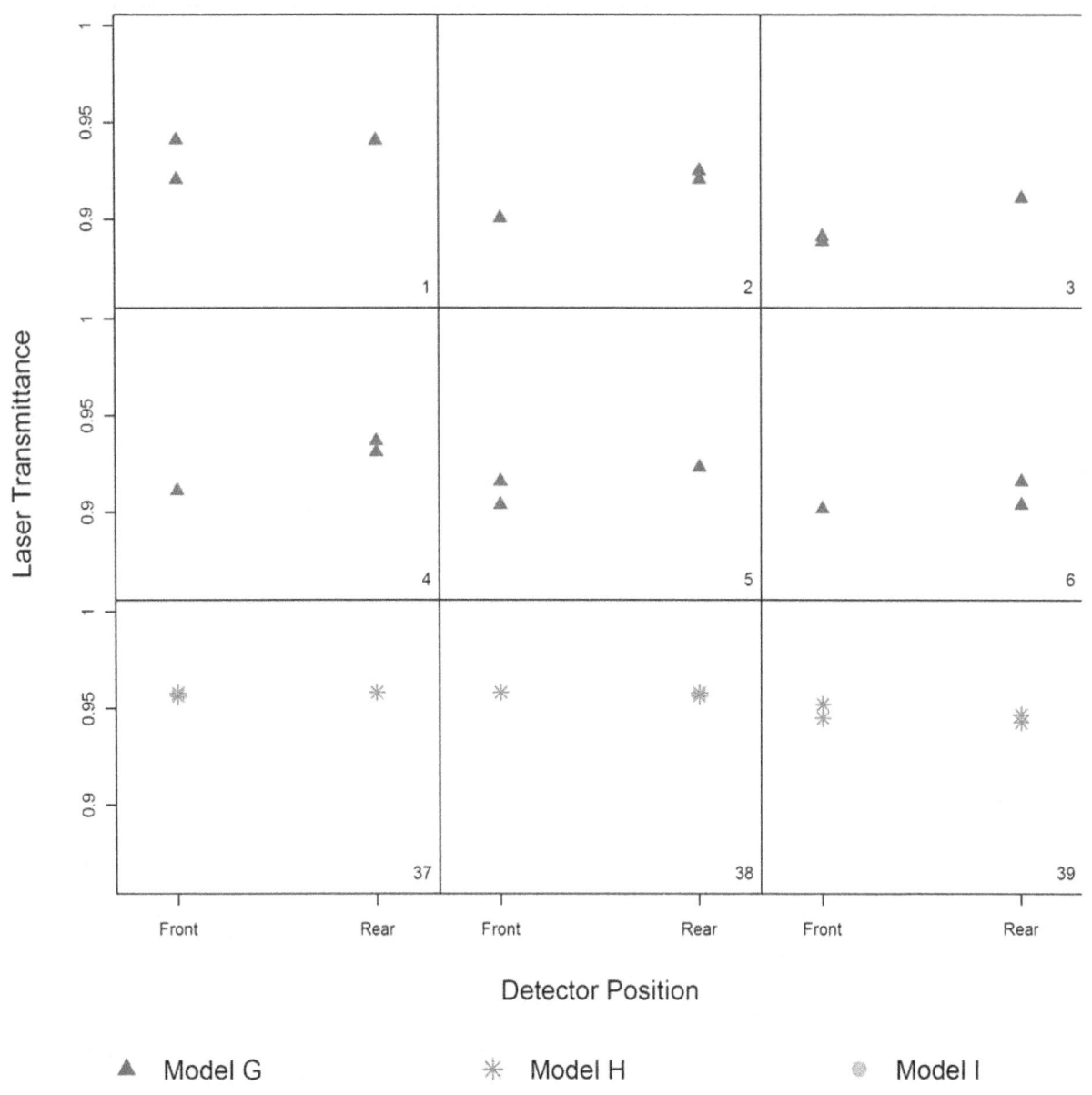

FIGURE H. 5: SET 3 PHOTOELECTRIC SMOKE ALARM RESPONSES VS. POSITION BY UNIT

The smoke alarm models are indicated by the color coding, and the unit identification numbers are shown in the lower right corner of each plot. These plots provide more detail about individual unit responses to position in the FEDE test apparatus. In all cases, the responses for each unit look relatively consistent, regardless of position.

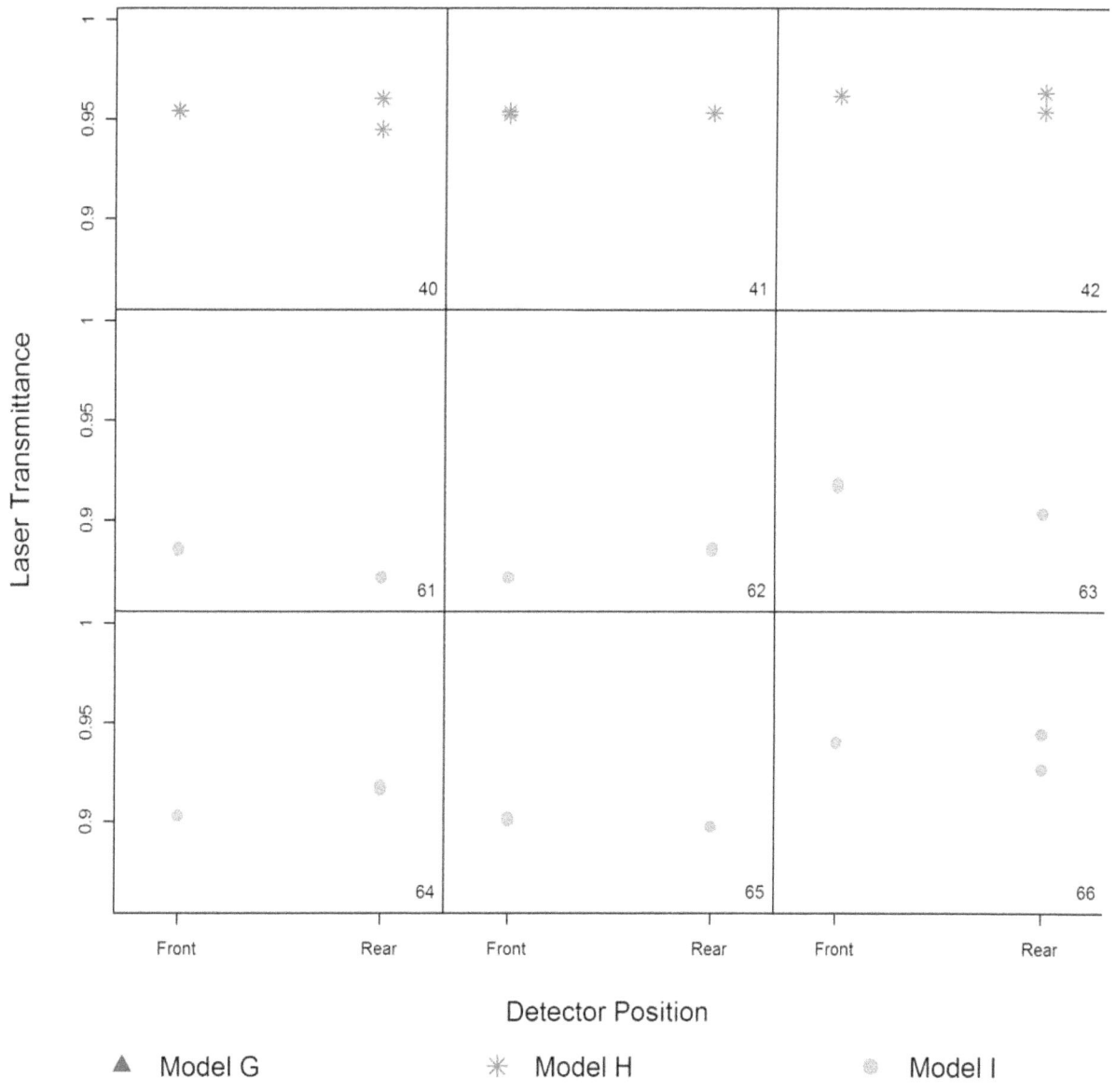

FIGURE H.5 CONTINUED: CAPTION AND INTERPRETATION ON PREVIOUS PAGE.

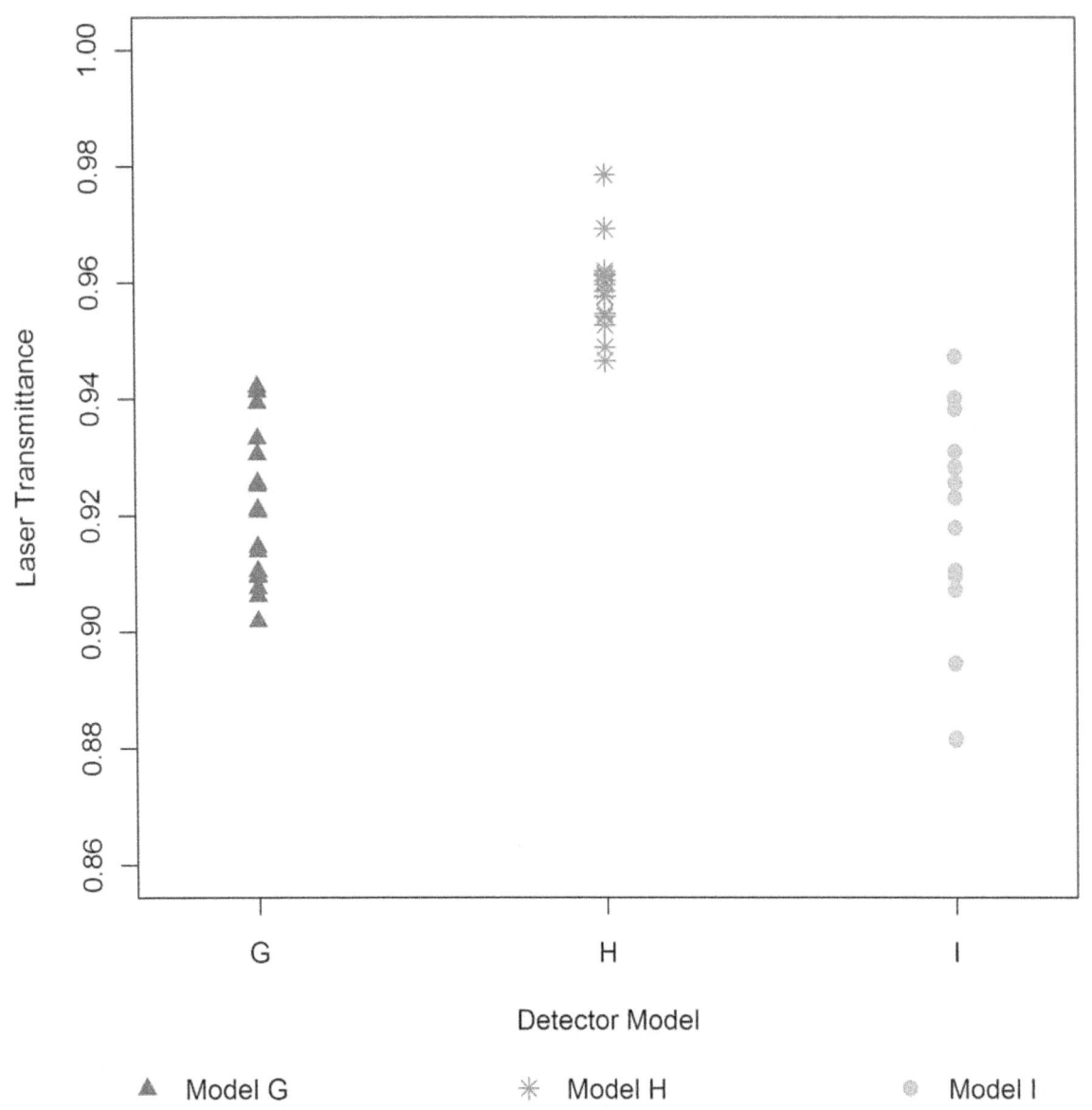

FIGURE I. 1: SMOKE ALARM RESPONSES VS. MODEL FOR SET 4 PHOTOELECTRIC SMOKE ALARMS

The responses from multiple tests of six different smoke alarms from each of three models are color coded by model. The model-to-model variation seen here is similar to that seen when the units were new, indicating that the accelerated aging did not impact relative unit responses on average. The scatter in the data looks essentially like the scatter observed when the units were new as well.

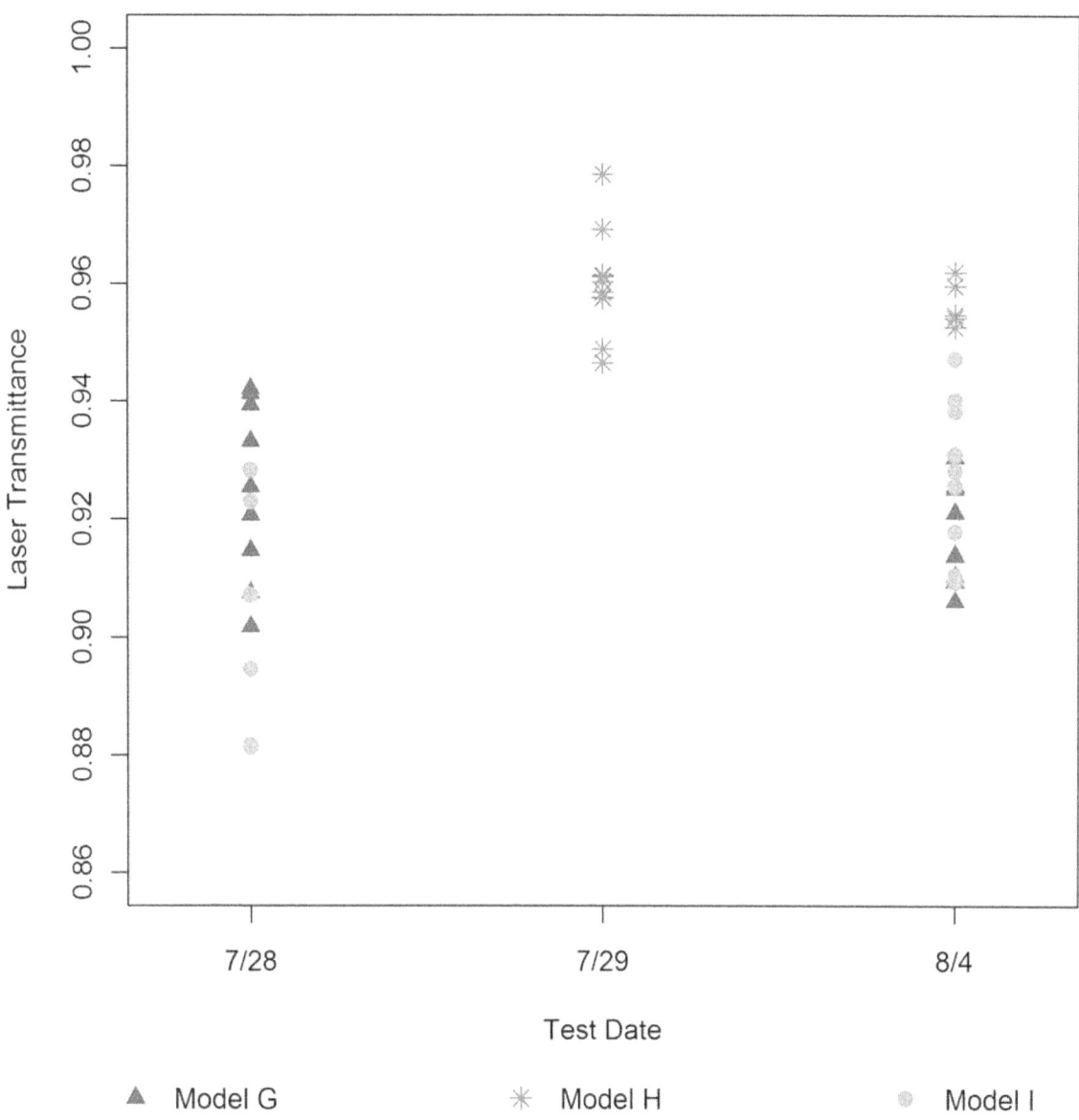

FIGURE I. 2: SET 4 PHOTOELECTRIC SMOKE ALARM RESPONSES VS. TEST DATE

The color coding indicates the smoke alarm model. The pattern in the data seen here, which suggests variation between days on which the different tests were performed, is again a reflection of model-to-model variations in the responses. This can be seen by noting the high responses for Model H units on both July 29 and August 4, along with the similarity of the readings for the Model G and Model I units on July 28 and August 4. The fact that only Model H units were measured on July 29, and that the Model H units have the highest mean and tightest scatter, accentuate the specious appearance of a day-to-day effect.

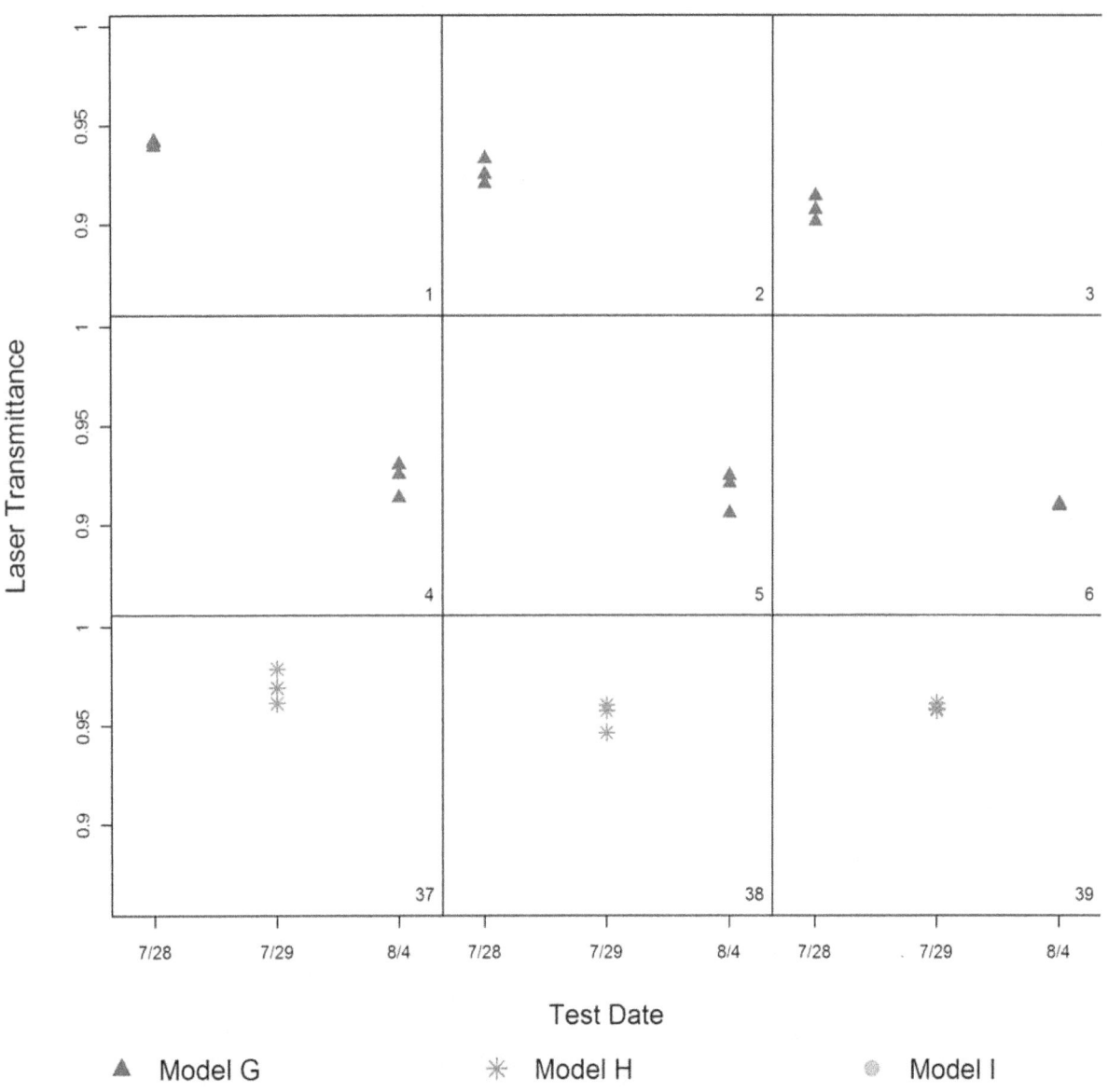

FIGURE I. 3: SET 4 PHOTOELECTRIC SMOKE ALARM RESPONSE VS. DATE BY UNIT

The smoke alarm models are indicated by symbol color. The unit identification numbers are given in the lower right corner of each plot. In this case, each unit was measured on only one day (and in one position in the FEDE). Looking at all six plots for each model, the responses for the different units look quite consistent overall. As when these smoke alarms were new, the differences in the models can be seen in absolute response levels, when comparing individual units as well.

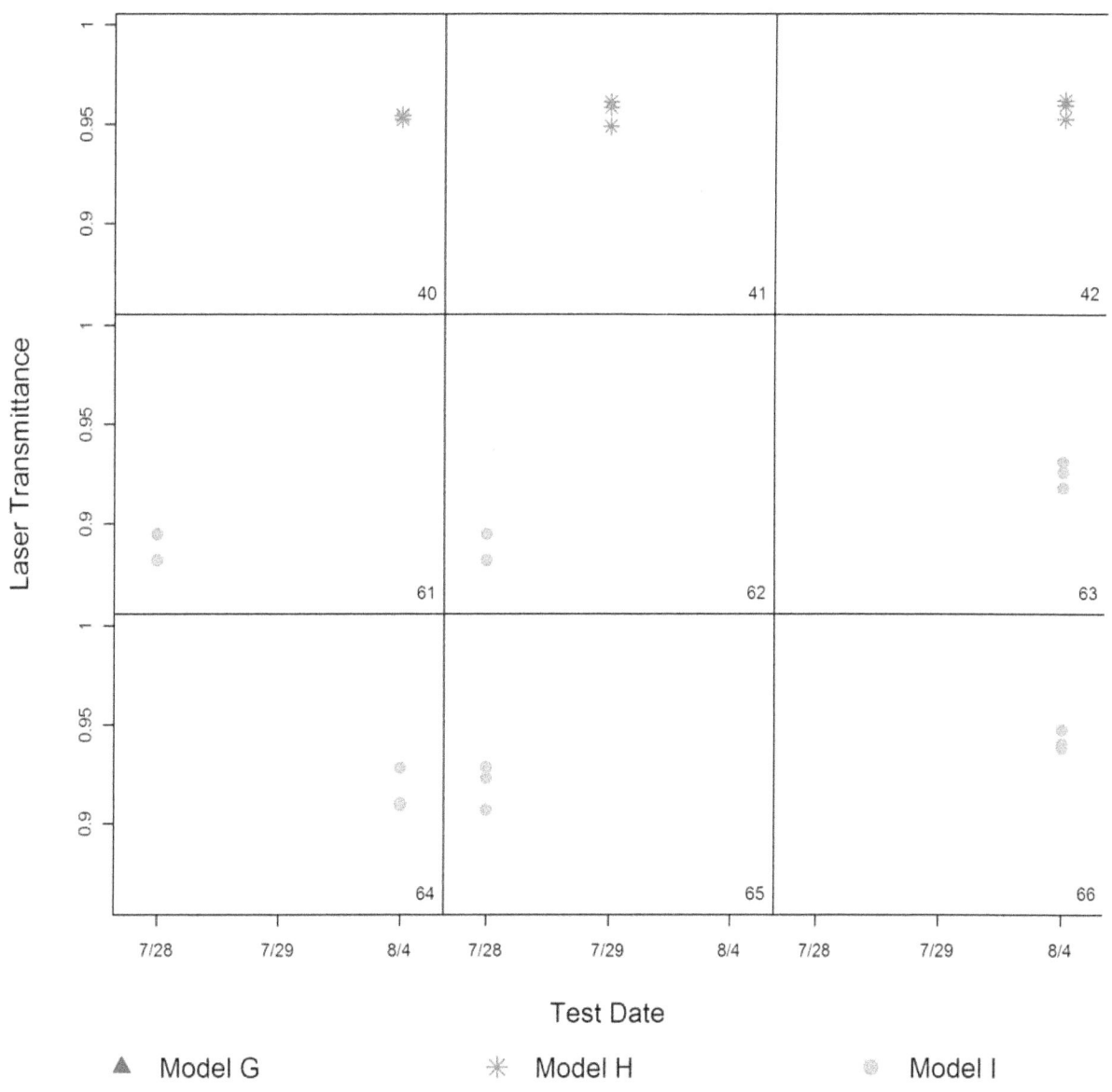

FIGURE I.3 CONTINUED: CAPTION AND INTERPRETATION ON PREVIOUS PAGE.

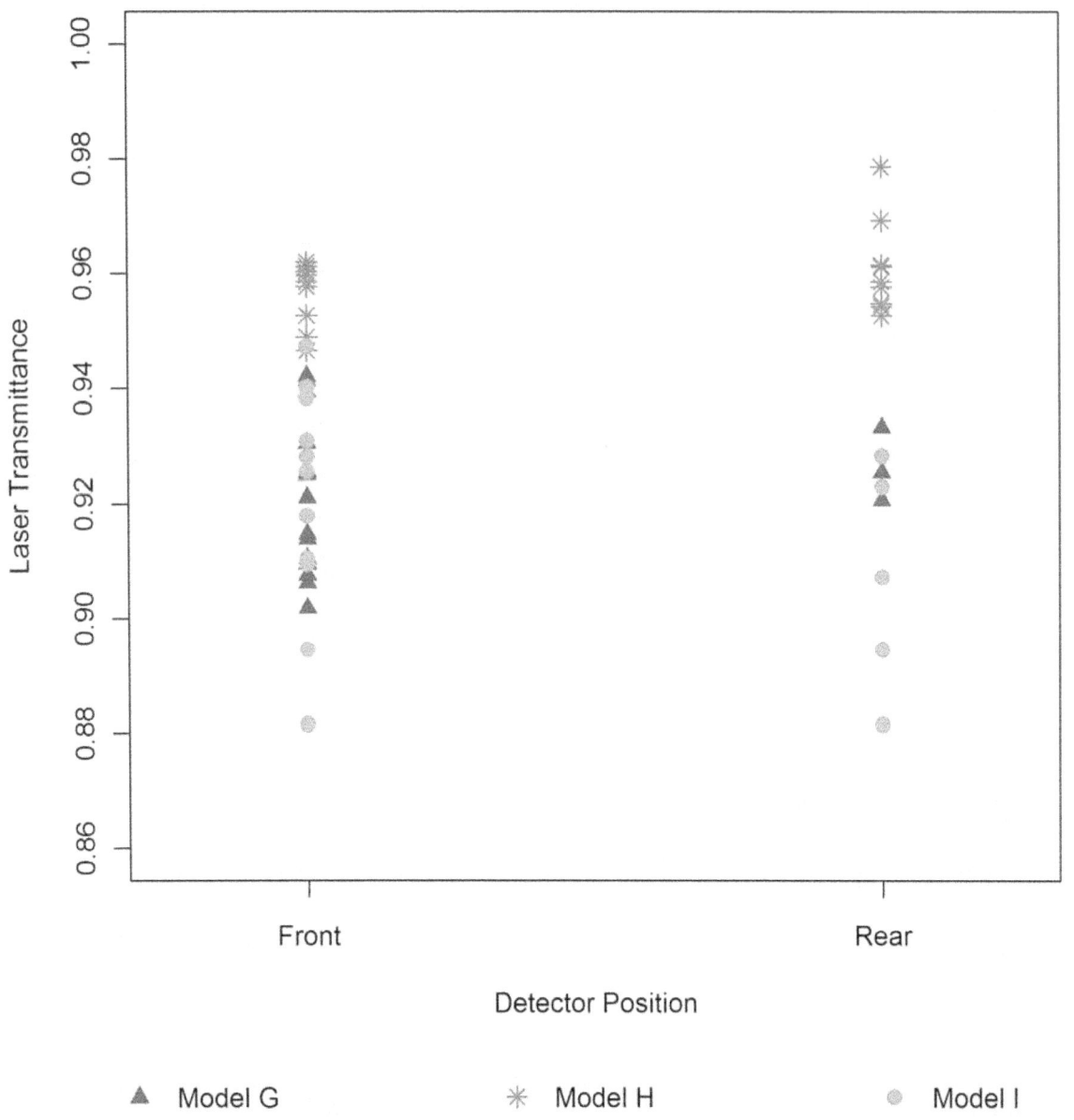

FIGURE I. 4: SET 4 PHOTOELECTRIC SMOKE ALARM RESPONSES VS. POSITION IN THE FEDE

The smoke alarm model is indicated by the color coding. The similar responses for each position indicate that this factor does not affect the results on average for any of the three models.

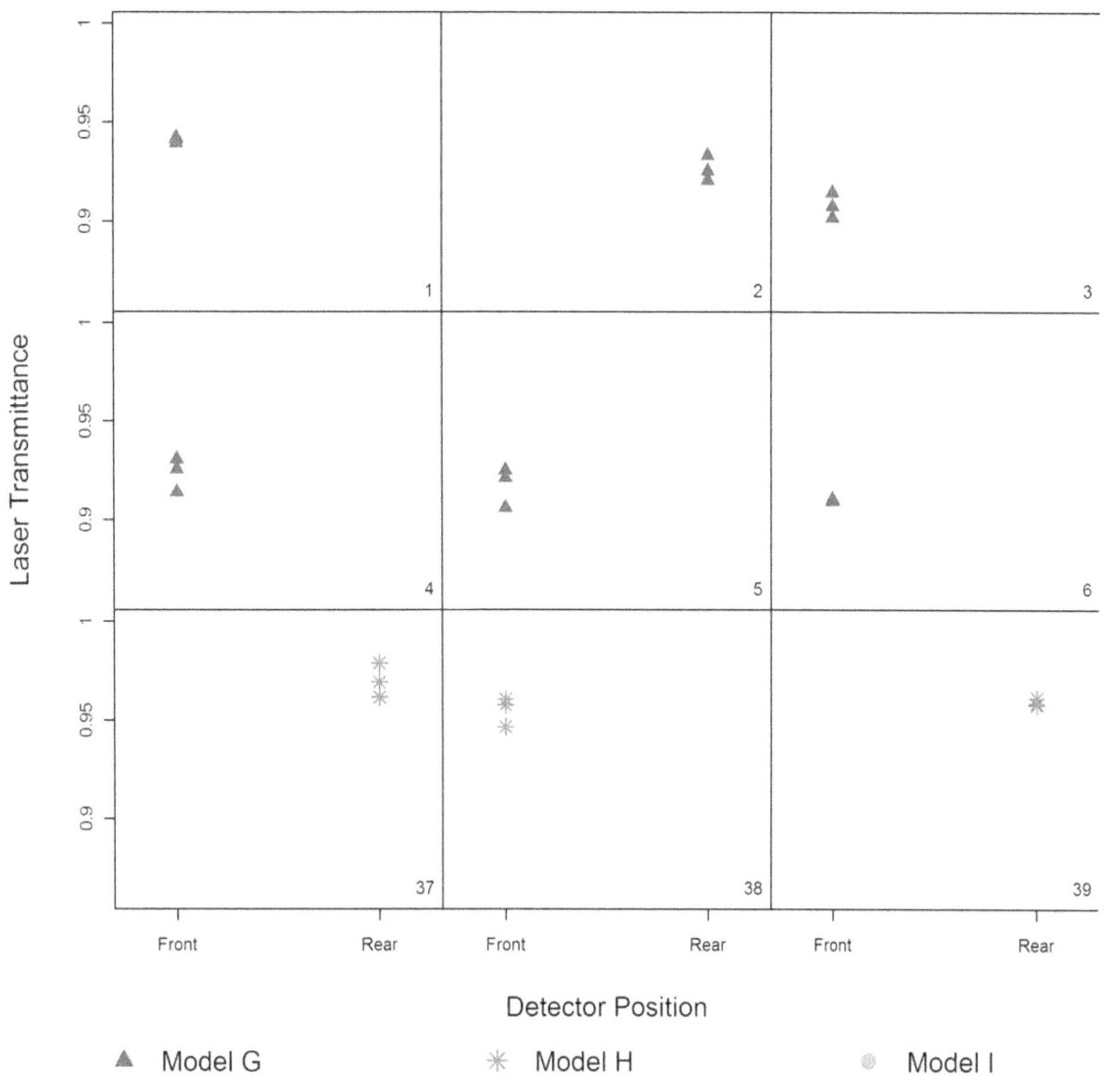

FIGURE I. 5: SMOKE ALARM RESPONSES VS. POSITION BY UNIT

The smoke alarm models are indicated by the color coding, and the unit identification numbers are shown in the lower right corner of each plot. These plots provide more detail about individual unit responses and show that each smoke alarm was only measured in one position after aging. In all cases, the responses across units look relatively consistent for each model of smoke alarms, regardless of position.

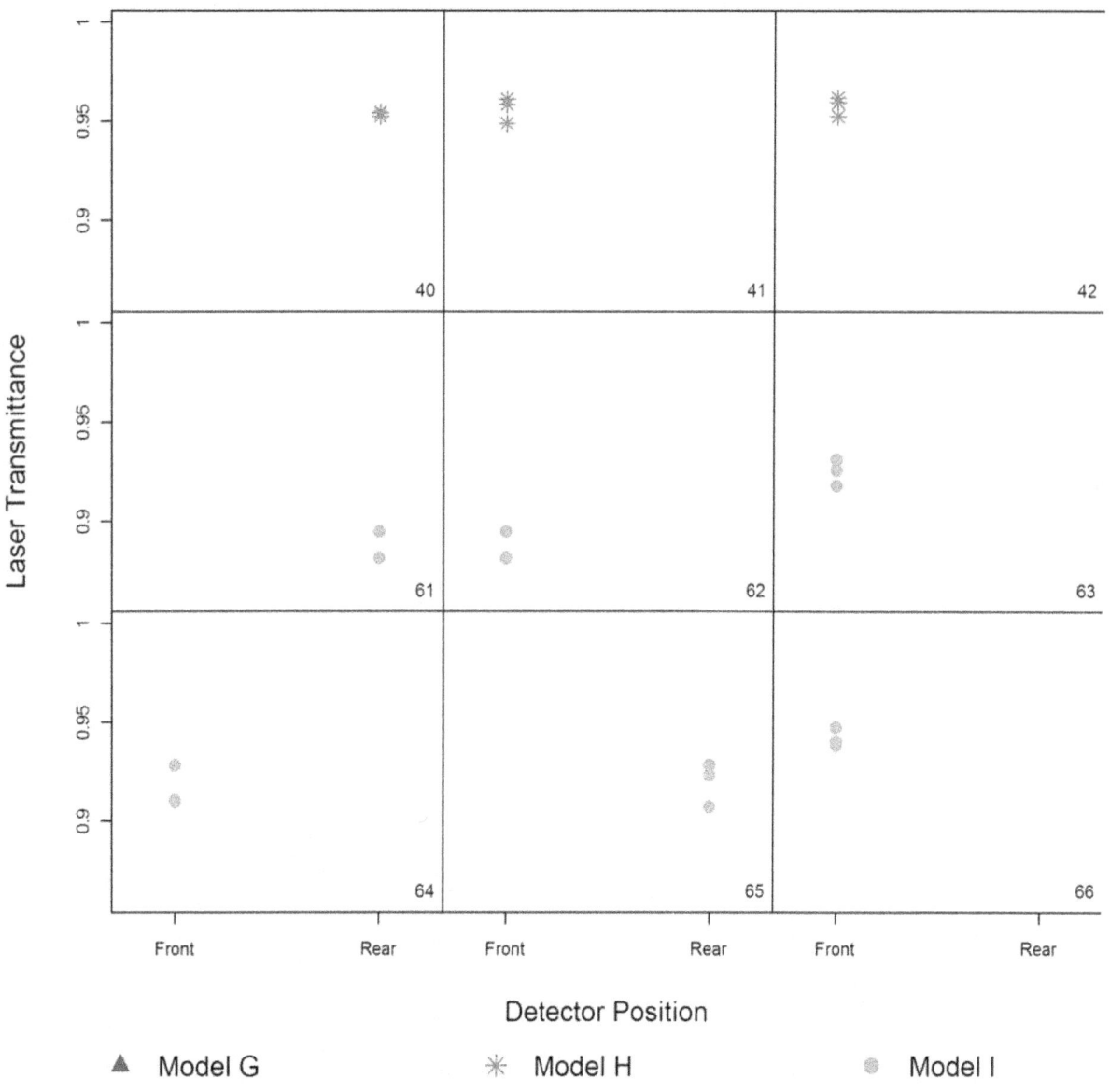

FIGURE I.5 CONTINUED: CAPTION AND INTERPRETATION ON PREVIOUS PAGE

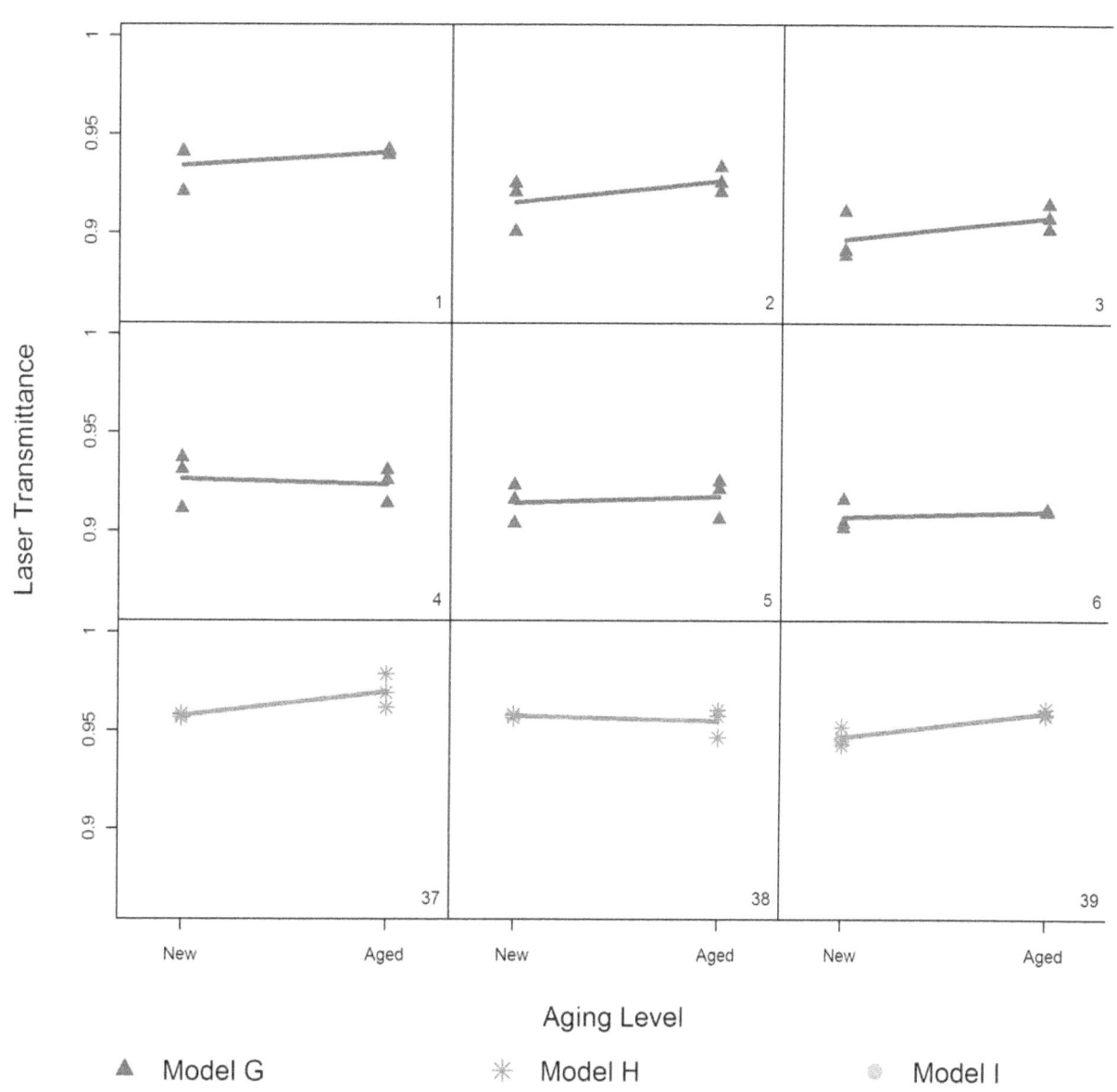

FIGURE J. 1: SMOKE ALARM RESPONSES FOR SET 3 VS. SET 4 PHOTOELECTRIC SMOKE ALARMS BY UNIT

The responses from multiple runs on six different smoke alarms from each of three models are color coded by smoke alarm model. The unit identification numbers are shown in the lower right corner of each plot. The diagonal lines on each plot, which connect the means of the runs before and after aging, facilitate determining whether the observed mean response for each unit increased or decreased after aging. These plots show that for Models G and I, most or all average smoke alarm reponses were higher after aging, than when new. For the Model H units, however, this effect appears to be a little weaker, with relatively large increases for only two of the units and two showing a decreased response after aging.

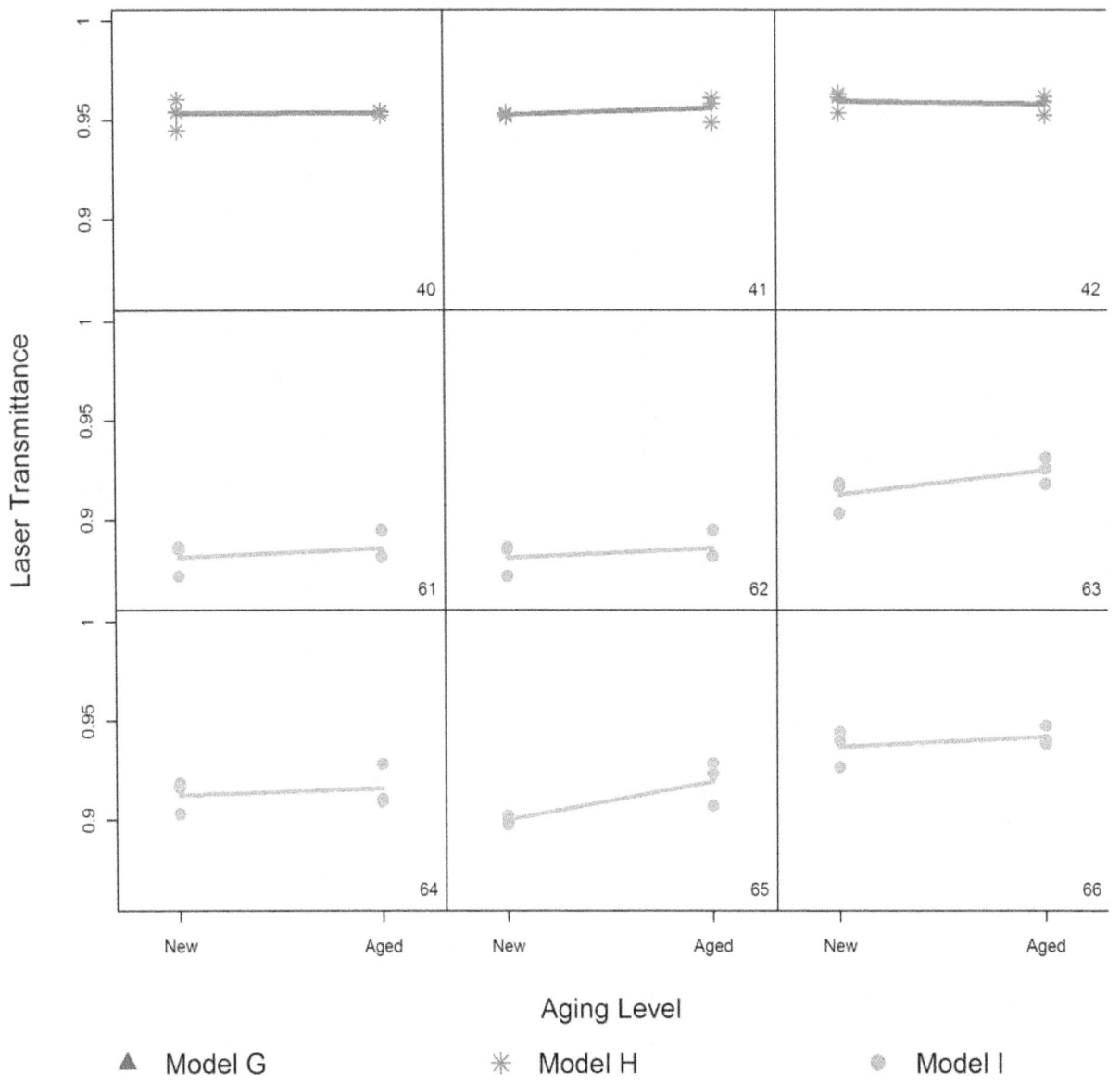

FIGURE J.1 CONTINUED: CAPTION AND INTERPRETATION ON PREVIOUS PAGE

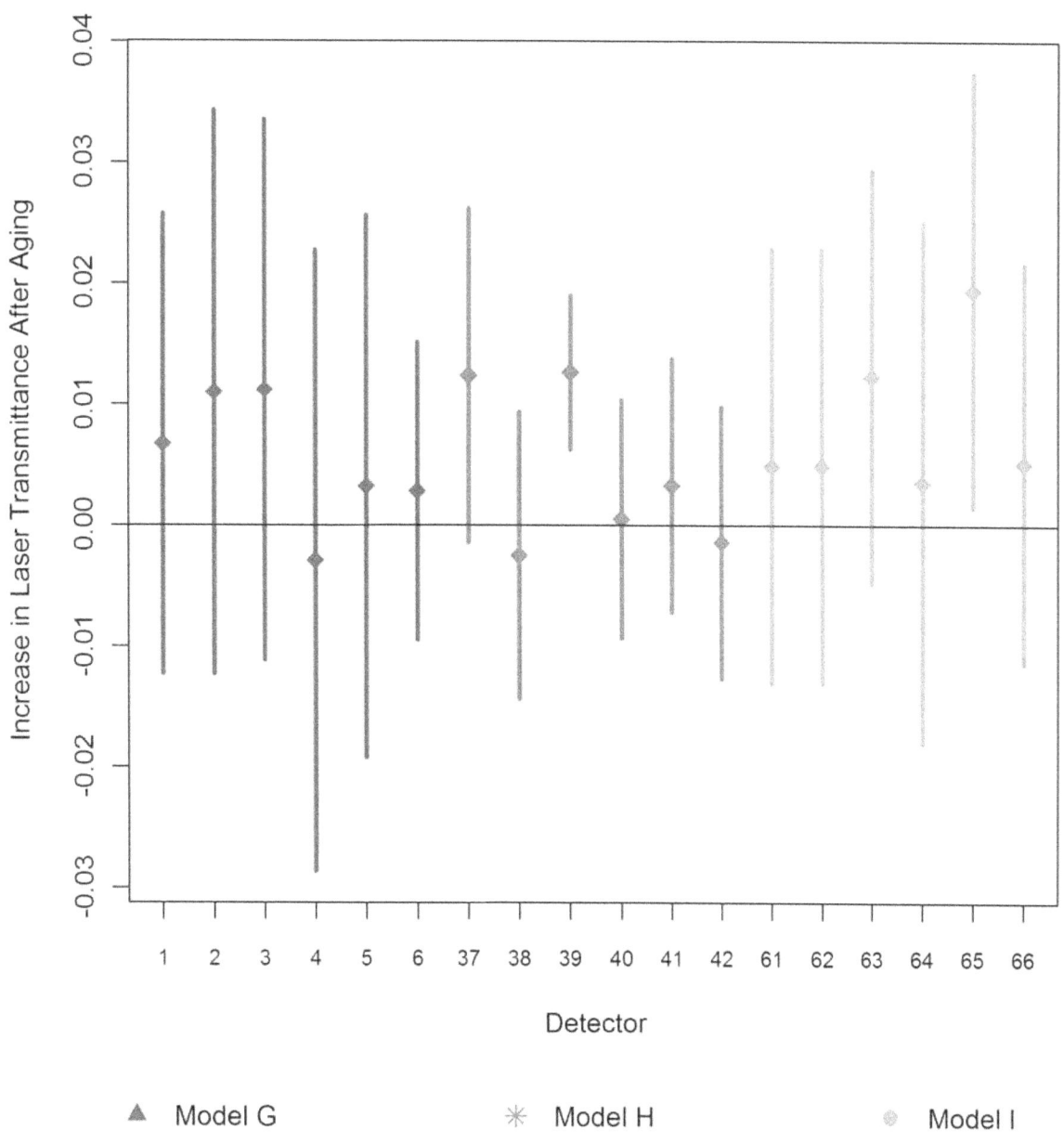

FIGURE J. 2: INDIVIDUAL CONFIDENCE INTERVALS (CI'S) FOR THE AVERAGE DIFFERENCE IN RESPONSE BETWEEN SET 3 AND SET 4 PHOTOELECTRIC SMOKE ALARMS BY UNIT

The results are color coded by smoke alarm model. Significant differences in response before and after aging are indicated by intervals that do not contain zero (*e.g.*, units 39 and 65). The fact that a difference in response for any individual unit is not significant does not mean there is no difference, just that any difference that exists cannot be detected with the amount of data for that unit. When results are pooled across units, differences in response before and after aging may be revealed.

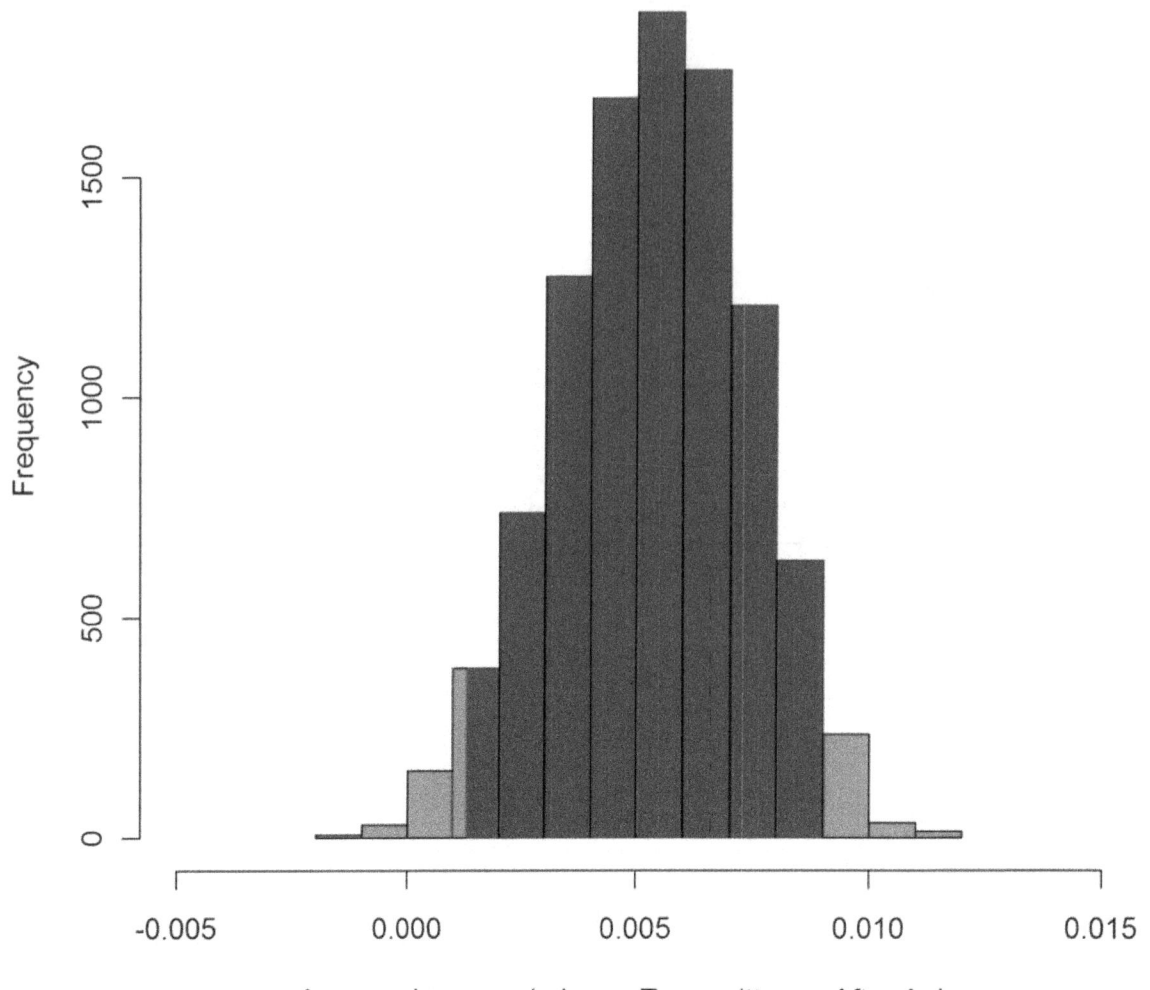

95 % Confidence Interval for Average Laser Transmittance Increase: (0.0013 , 0.009)

FIGURE J. 3: HISTOGRAM SHOWING THE DISTRIBUTION OF RESAMPLED ESTIMATES OF THE AVERAGE
DIFFERENCE IN SMOKE ALARM RESPONSE FOR ALL MODEL G PHOTOELECTRIC SMOKE ALARMS

The blue portion of the histogram shows the central 95 % of the distribution, while the red portions
in the tails of the distribution (and all more extreme values) comprise the least likely 5 % of the
distribution (2.5 % in each tail). The fact that a difference of zero falls slightly outside the central
95 % bounds indicates that the average difference in unit response before and after aging is barely
statistically significant. More information on the methodology used to compute this confidence
interval is given in Appendix A.

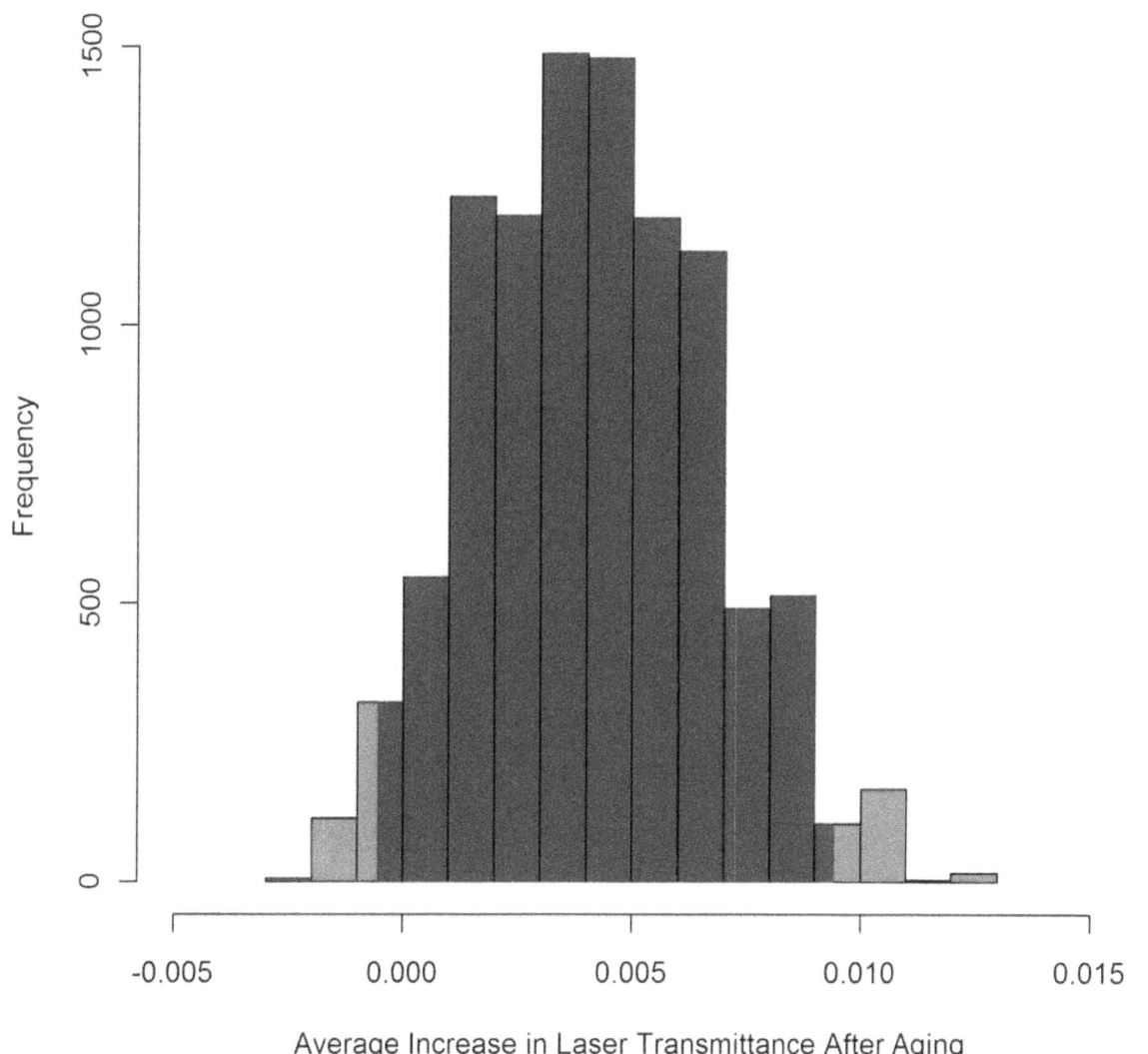

Average Increase in Laser Transmittance After Aging

95 % Confidence Interval for Average Laser Transmittance Increase: (-0.00055 , 0.0094

FIGURE J. 4: HISTOGRAM SHOWING THE DISTRIBUTION OF RESAMPLED ESTIMATES OF THE AVERAGE
DIFFERENCE IN SMOKE ALARM RESPONSE FOR ALL MODEL H PHOTOELECTRIC SMOKE ALARMS

The blue portion of the histogram shows the central 95 % of the distribution, while the red portions
in the tails of the distribution (and all more extreme values) comprise the least likely 5 % of the
distribution (2.5 % in each tail). The fact that a difference of zero falls within the central 95 %
bounds indicates that the average difference in unit response before and after aging is not
statistically significant. More information on the methodology used to compute this type of
confidence interval is given in Appendix A.

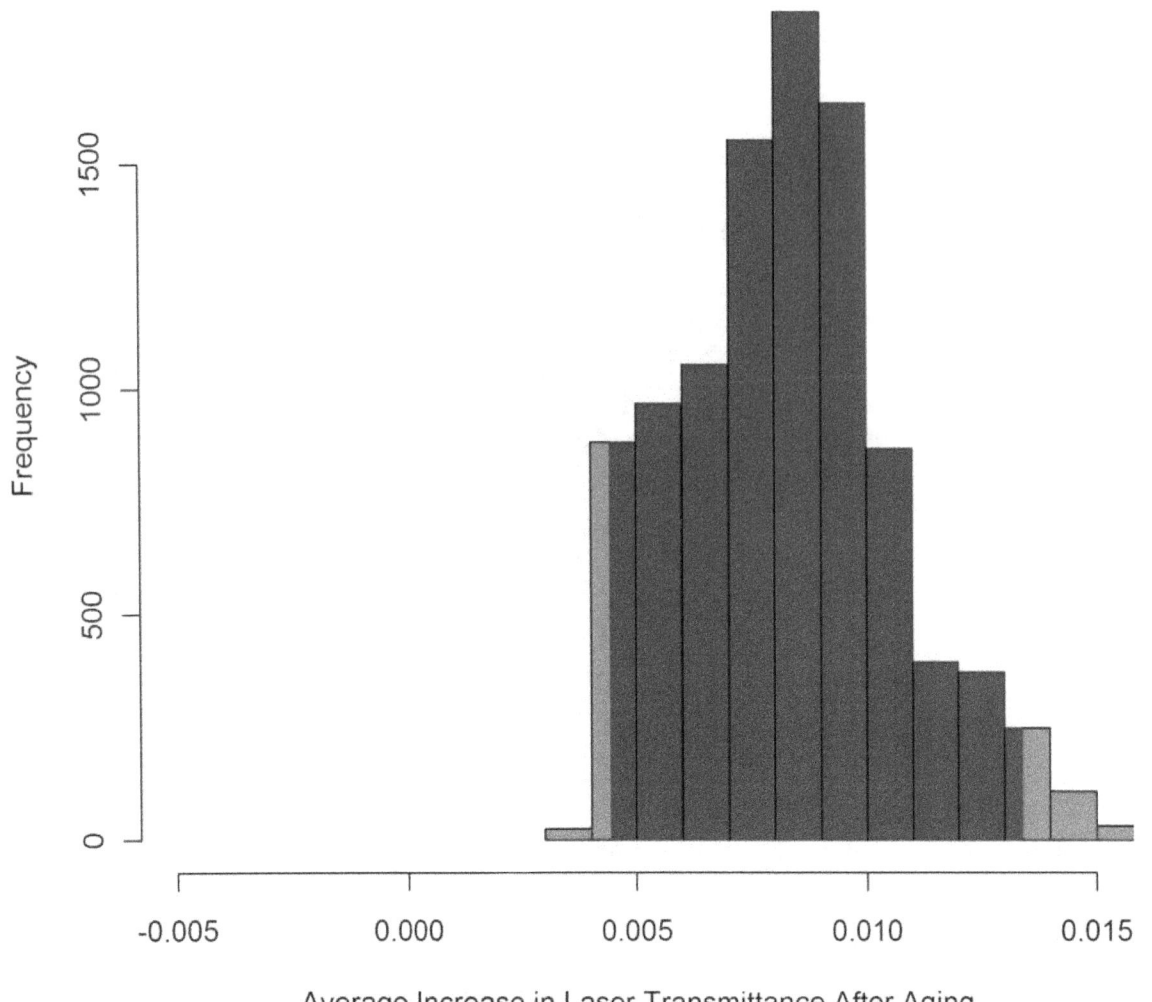

95 % Confidence Interval for Average Laser Transmittance Increase: (0.0045 , 0.013)

FIGURE J. 5: HISTOGRAM SHOWING THE DISTRIBUTION OF RESAMPLED ESTIMATES OF THE AVERAGE
DIFFERENCE IN SMOKE ALARM RESPONSE FOR ALL MODEL I PHOTOELECTRIC SMOKE ALARMS

The blue portion of the histogram shows the central 95 % of the distribution, while the red portions
in the tails of the distribution (and all more extreme values) comprise the least likely 5 % of the
distribution (2.5 % in each tail). In this case, zero does not fall within the central 95 % bounds,
which indicates that the average difference in unit response before and after aging is barely
statistically significant. More information on the methodology used to compute this type of
confidence interval is given in Appendix A.

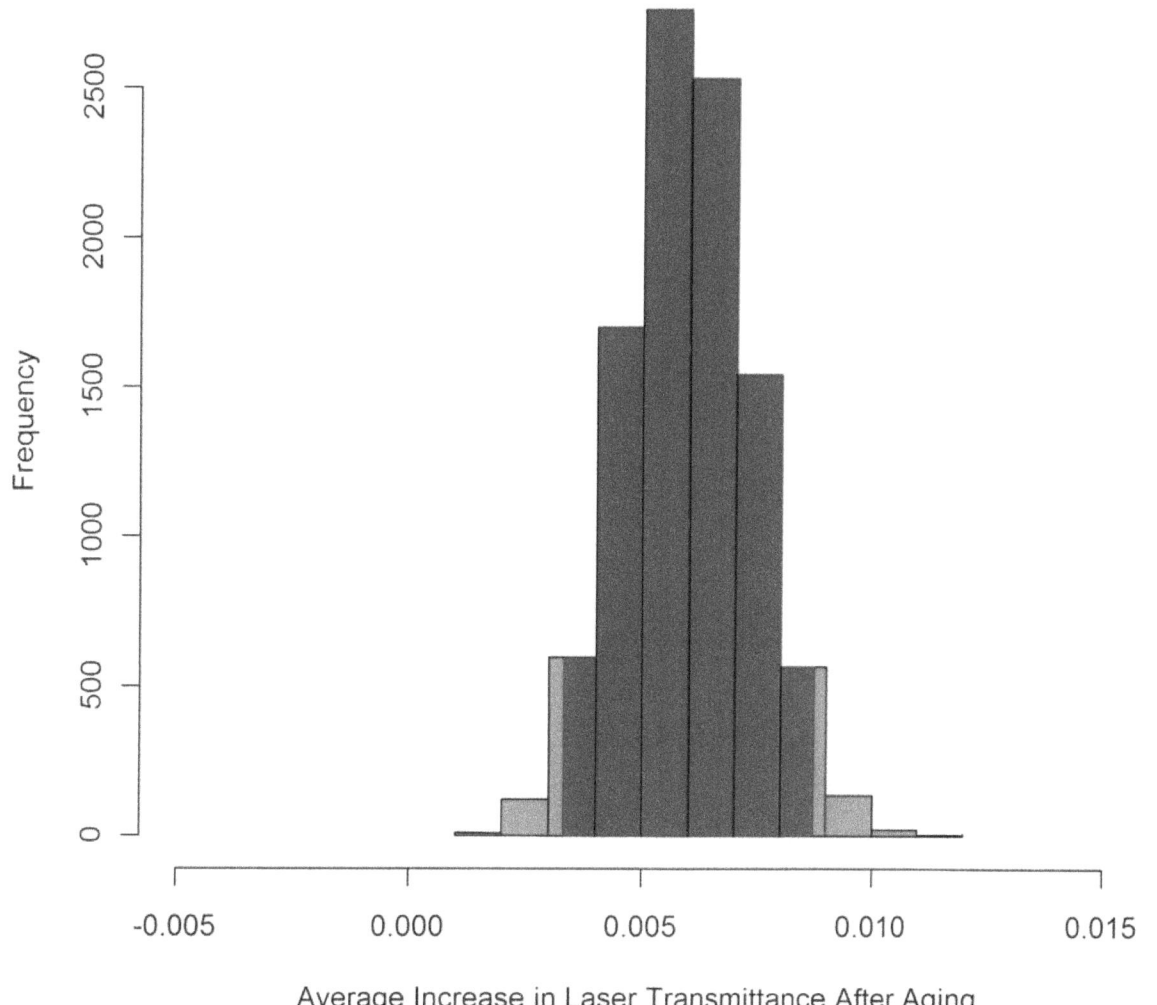

Average Increase in Laser Transmittance After Aging

95 % Confidence Interval for Average Laser Transmittance Increase: (0.0033 , 0.0087)

FIGURE J. 6: HISTOGRAM SHOWING THE DISTRIBUTION OF RESAMPLED ESTIMATES OF THE AVERAGE
DIFFERENCE IN SMOKE ALARM RESPONSE FOR ALL MODELS OF PHOTOELECTRIC SMOKE ALARMS

This histogram assumes that the distribution of different smoke alarms in use is the same as the distribution of smoke alarm models sampled (approximately 50 % Model A and 25 % each of the other two models). The blue portion of the histogram shows the central 95 % of the distribution, while the red portions in the tails of the distribution (and all more extreme values) comprise the least likely 5 % of the distribution (2.5 % in each tail). The fact that a difference of zero falls just outside the central 95 % bounds indicates that the average difference in unit response before and after aging is barely statistically significant. More information on the methodology used to compute this confidence interval is given in Appendix A.

REFERENCES

1. http://www.cpsc.gov/info/drywall

2. Environmental Health & Engineering, Inc. "Final Report on an Indoor Environmental Quality Assessment of Residences Containing Chinese Drywall." EH&E Report 16512, submitted to the Consumer Product Safety Commission, Bethesda, MD, January 28, 2010.

3. Maddalena, R., Marion, R., Moya M., and Apte, M.G., "Small-Chamber Measurements of Chemical Specific Emission Factors for Drywall." Report Number: LBNL-3986E, Ernest Orlando Lawrence Berkeley National Laboratory, Berkeley, CA, October 2010.

4. Abbott, W., "The Development and Performance Characteristics of Mixed Flowing Gas Test Environment," *IEEE Transactions on Components, Hybrids, and Manufacturing Technology* **11**, 22-35, (1988).

5. Glass, S. J., Mowry, C. D., and Sorensen, N. R. "Report on Accelerated Corrosion Studies of Electrical Components." Sandia Report SAND2011-1539. Sandia National Laboratories, Albuquerque, NM, March 2011.

6. *UL 217: Single and Multiple Station Smoke Alarms*, Underwriters Laboratories. Northbrook, IL, (2008).

7. Cleary, T. G.; Donnelly, M. K.; Grosshandler, W. L. "The Fire Emulator/Detector Evaluator," Proceedings, International Conference on Automatic Fire Detection: AUBE '01, National Institute of Standards and Technology. March 25-28, 2001, Gaithersburg, MD, Beall, K.; Grosshandler, W. L.; Luck, H., Editors, pp. 312–323, (2001).

8. Bukowski, R.W., et al., "Performance of Home Smoke Alarms Analysis of the Response of Several Available Technologies in Residential Fire Settings," NIST Technical Note 1455-1, National Institute of Standards and Technology, Gaithersburg, MD, 396 pages, (2008).

9. Cleary, T. G.; Chernovsky, A.; Grosshandler, W. L.; Anderson, M., "Particulate Entry Lag in Smoke Alarms," National Institute of Standards and Technology. Annual Conference on Fire Research: Book of Abstracts, pp. 11-12, Beall, K. A., Editor(s), National Institute of Standards and Technology, Gaithersburg, MD, (1998). http://fire.nist.gov/bfrlpubs/fire98/PDF/f98094.pdf

10. Efron, B. and Tibsirani, R. J. , *An Introduction to the Bootstrap*, vol. 57 of *Monographs of Statistics and Applied Probability*, Chapman and Hall, Boca Raton, FL (1993).

11. Ryan, T. P., *Modern Engineering Statistics*, John Wiley and Sons, Hoboken, NJ (2007).